Contemporary Urban Design Thinking

Series editor
Rob Roggema
Faculty of Design Architecture Building
University of Technology Sydney
Broadway, NSW, Australia

Cittaideale
Wageningen, The Netherlands

This series will investigate contemporary insights in urban design theory and practice. Urbanism has considerably changed and developed over the years and is about to undergo a transformation moving into a new era.

In the 1990's and early 2000's economic driven urban design was prevalent in many countries around the world. Moving forward it is no longer feasible to continue to develop in the same way and new ideas for creating urbanism are urgently required.

This series will publish titles dealing with innovative methods of urbanism including, sustainability driven urbanism, smart urbanism, population driven urbanism, and landscape based urban design.

The series will include books by top researchers and leaders in the fields of urban design, city development and landscape urbanism. The books will contain the most recent insights into urbanism and will provide actual and timely reports filling a gap in the current literature.

The series will appeal to urbanists, landscape architects, architects, policy makers, city/urban planners, urban designers/researchers, and to all of those interested in a wide-ranging overview of contemporary urban design innovations in the field.

More information about this series at http://www.springer.com/series/15794

Rob Roggema
Editor

Contemporary Urban Design Thinking

The Australian Approach

Springer

Editor
Rob Roggema
Faculty of Design Architecture Building
University of Technology Sydney
Broadway, NSW, Australia

Cittaideale
Wageningen, The Netherlands

ISSN 2522-8404 ISSN 2522-8412 (electronic)
Contemporary Urban Design Thinking
ISBN 978-3-030-06335-1 ISBN 978-3-319-91950-8 (eBook)
https://doi.org/10.1007/978-3-319-91950-8

Printed on acid-free paper

This Springer imprint is published by the registered company Springer Nature Switzerland AG
The registered company address is: Gewerbestrasse 11, 6330 Cham, Switzerland

Contents

About the Contributors

Craig Allchin is an architect and urban planner, director of Six Degrees Urban and Adjunct Professor of Architecture at UTS, Sydney. He has worked on the city across a range of scales. In the 1990's he lived in the Melbourne city center and worked as a founding partner of Six Degrees, on urban renewal projects as part of the revitalisation of the Melbourne City Centre. He moved to Sydney and spent four years travelling to China working on over 40 district and city master plans, then four years as Director of Urban Design for the NSW Government's 25 Year Strategic Metropolitan Plan for Sydney.

Peter Bishop is Professor of Urban Design at The Bartlett School of Architecture, University College London and a Distinguished Visiting Scholar at the University of Technology Sydney. He is also a director at the London based architects practice Allies and Morrison. Before these roles he was Director of Design for London, the Mayor of London's architectural design agency.

Andy van den Dobbelsteen is professor of Climate Design & Sustainability at TU Delft and principal investigator at AMS Institute, amongst other functions. He teaches students sustainable building design and conducts research related to energy transition, climate adaptation and circular cities.

Hugh Gardner is member of Arup's Integrated Cities Planning Team in Sydney and has devoted his career to developing a broad and integrated set of skills that can be applied to the design of healthy, sustainable, equitable and socially balanced urban environments. Highlights have included three years working with UrbanGrowth NSW to develop a framework for 50 years of renewal activity at Central to Eveleigh, two years working with the City of Sydney on their digital strategy, developing a city vision with all the directorates in Canberra.

Justine Kinch leads the Design + planning team in NSW at AECOM and has recently taken the role as Director for Western Sydney. In this role she is responsible for building brand and a culture of collaboration across AECOM's multidisciplinary teams to get better outcomes for Western Sydney. One of her key focus areas is to develop Thought Leadership on how to stage and deliver optimised enabling infrastructure for the new western city. Justine is an active member of the industry and is currently chairing the Urban Land Institutes Women's Leadership Initiative Committee and the State Education Committee for Australian Institute of Landscape Architects.

Gerard Reinmuth is Professor of Practice at the School of Architecture in the University of Technology, Sydney. He is also Founding Director and Principal of the architectural practice TERROIR. His interest in the role of the architect in city-making is explored through this combination of research and practice roles and the interplay between the experiences gained in each.

Rob Roggema is Professor of Sustainable Urban Environments in the Faculty of Design, Architecture and Building at the University of Technology Sydney (UTS). He is an international renowned design expert on sustainable urbanism, climate adaptation, renewable energy landscapes and the design of urban agriculture. He has held positions at several universities in the Netherlands and Australia, state and municipal governments and design consultancies.

Barbara Schaffer is Principal Landscape Architect in the Government Architects Office since 2006. With over twenty years of professional experience Barbara has a strong track record in place-making, strategic planning, design and the delivery of complex high profile public places, within both government and the private sector. Her projects have received numerous awards including the Australian Institute of Landscape Architects's Award's Award of Excellence for Outstanding Achievement in Landscape Architecture and the National AILA Award for Environmental Planning.

Roderick Simpson is Environment Commissioner for the Greater Sydney Commission. He is an architect and urban designer, and an Associate Professor at University of Technology Sydney. He has worked for the Commonwealth and provided consultancy services to State and local governments in planning and urban design. He led urban design for the City of Sydney Sustainable Sydney 2030 strategy with the SGS consortium. He is a member of various housing, transport and architectural design review panels.

Jay Stricker is an Industry Director in Transportation for AECOM and has led major transportation studies both in AECOM and previously in her senior management roles in the former NSW Roads and Traffic Authority, now Roads and Maritime Services. She has more than 30 years of experience in senior leadership roles in the public sector primarily focussed on the planning and delivery of major water and transport infrastructure projects. Jay has substantial experience in the planning and evaluation of highway projects and was closely involved in the development of the Pacific Highway Upgrade Program from late 1995. In 2008 she joined AECOM and has led a variety of multi-disciplinary transport related studies. Jay is the Deputy Chair of the Roads Australia Sustainability Policy Chapter with a focus on resilience of cities and regions.

Georgia Vitale is an experienced city planner, urban strategist and renewal specialist with experience gained in projects throughout Australasia, Sri Lanka, the Middle East and the UK. Georgia leads Arup's Integrated Cities Planning Team in Sydney. Georgia's city resilience experience includes working in Sri Lanka post the Asian Tsunami, supporting the Greater Sydney Commission's resilience planning and most recently with the Asian Development Bank and the Rockefeller Foundation as part of the Urban Climate Change Resilience Trust Fund project as a subject matter expert on cities.

Chapter 1
Australian Urbanism; State of the Art

Rob Roggema

Abstract This book on Contemporary Urban Design Thinking focuses on the Australian approach. Australia, often compared with the US in the ways its cities have grown, has its own particular way cities are developed. Though the cities are car-based, even car-dependent, and landownership is a dominant in the development process, the Australian cities are also characterised by a strong community sense at the neighbourhood level. At the same time, Urbanism and Urban Design are disciplines that are somewhat under the radar in the design education in the country, still thought leadership on the design of cities is provided through urbanists, originally educated as landscape architects or architects. In this first volume of the series 'Contemporary Urban Design Thinking, the Australian Approach', illuminates these cutting edge visions and practices from a suite of perspectives.

1.1 Introduction

This book on Contemporary Urban Design Thinking focuses on the Australian approach. Australia, often compared with the US in the ways its cities have grown, has its own particular way cities are developed. Though the cities are car-based, even car-dependent, and landownership is a dominant in the development process, the Australian cities are also characterised by a strong community sense at the neighbourhood level. At the same time, Urbanism and Urban Design are disciplines that are somewhat under the radar in the design education in the country, still thought leadership on the design of cities is provided through urbanists, originally educated as landscape architects or architects. In this first volume of the series 'Contemporary Urban Design Thinking, the Australian Approach', illuminates these cutting edge visions and practices from a suite of perspectives.

In every part of the world urbanism takes different shapes, has different regulations and planning systems (Albrechts 2004; Alterman 2001; Hall 1992; Popper

R. Roggema (✉)
Faculty of Design Architecture Building, University of Technology Sydney, Broadway, NSW, Australia

Cittaideale, Wageningen, The Netherlands
e-mail: rob@cittaideale.eu

R. Roggema (ed.), *Contemporary Urban Design Thinking*, Contemporary Urban Design Thinking, https://doi.org/10.1007/978-3-319-91950-8_1

Fig. 1.1 Sydney Opera House and Harbour Bridge. (photo: Shio Ang Yeong Hui)

1988). In the Netherlands (Van der Valk 2002; Hajer and Zonneveld 2000; Evers et al. 2000), Scandinavia (Hofstad 2013; Nilsson 2007; Næss 2007) and Germany (Schmidt and Buehler 2007; BBR 1993; Fuerst 2003; Kunzmann 2001) urban planning is strongly government driven and regulated, while many planning systems in developing countries are almost absent, inducing insurgent urban planning (Miraftab 2009; Holston 1995; Miraftab and Wills 2005; Miraftab et al. 2008; Roy 2005). Systems in the UK (Roberts and Lloyd 1999; Wannop 1995; Wong et al. 2000), USA (Kayden 2001; Popper 1988) and Australia (Freestone 2007; Gleeson and Low 2000; Orchard 1999; Randolph 2004; Toon and Falk 2003; Bunker 2009; Forster 2006; Gleeson 2000) are strongly market driven (Brenner et al. 2012; Križnik 2014) and decision-making taking place is strongly based on land ownership.

The Australian approach has led to a type of urbanisation that on the one hand is very successful, though on the other has its downsides. Australian cities have great city centres and are famous because of their iconic sites, such as the Sydney Opera House and Harbour Bridge in Sydney (Fig. 1.1). Year after year, Australian cities appear in the top of most liveable cities of the world, with Melbourne (1), Adelaide (5) and Perth (7) in 2017s top 10 (The economist Intelligence Unit, 2017), and Sydney just outside of this.

The flipside of being most liveable is several Australian cities, especially Sydney and Melbourne, are some of the most expensive cities to live in (Martin 2018; NUMBEO,2018). One of the effects of the ongoing rise in housing prices is a relatively low affordability (Irvine and Wade 2017; Troy and Van den Nouwelant 2015), and a city that grows towards cheaper options. These cheaper options are found through developing greenfield sites in low densities at a distance from the city. This sprawl continues to occupy new landscapes, in Sydney for instance towards the west (see Chaps. 2 and 3). Housing in Western Sydney, amongst other growth areas in Australian cities takes shape in the form of houses that are built as large as possible on the plot, leaving minimal spaces on the plot between neighbours and hardly pro-

Fig. 1.2 Middleton Grange, in Sydney's western suburbs (Google Earth)

viding a front or rear garden. The roof is covered with black materials, increasing the heat inside, reason why every home has to have an AC to deal with increasing heat in the future. The streets are broad, asphalted and not lined with trees, making the hot climate even harsher to cope with. Energy bills are rising and the distance to the CBD is ever growing, which also increases the costs of the families living in these neighbourhoods (Fig. 1.2).

At the same time the differences in housing prices shine light on a deeper underlying divide. The south-western part of the Sydney metropolitan region is structurally disadvantages from a social (Browne 2016), education level (Ting and Bagshaw 2016), health (Aubusson 2017), and income (Wade 2016) perspective (see also Chap. 7). These differences on either side of the so-called latte-line (Saulwick 2016) also known as the smashed avocado line cause a fundamental divide within the Sydney region.

As everywhere the decision-making in the political process (Hanssen, 2012) determines the outcomes in spatial reality (see Chaps. 4 and 5). In the Australian context landownership is a decisive factor in spatial decision-making (see Chap. 3). The perception of many decision-makers, politicians, bureaucrats, but also the general public believe that land cannot be developed in any other way than the current sprawling way, because landowners would then see their revenues reduced. If the future quality of an area herewith is increased or decreased depends totally on the landowners in question. Some have a long-term perspective in developing the land and take factors of longevity in account, while others only look at the short-term financial gain. This process of ownership-led future planning is under increasing scrutiny as the challenges in the (near) future become bigger and bigger. Fast population growth, economic pressure and climatic impacts restrain the options for short term gain, and imply the longer term quality of an area will determine whether the local population will survive and/or have the quality of life people demand.

1.2 Current Challenges

There lies a huge task ahead as the population of Australian cities (McCrindle 2015; Infrastructure Australia 2015) is expected to grow much faster (see Chap. 2) than London (Greater London Authority 2016) or New York (Regional Plan Association 2016) and even faster than most Chinese cities. Sydney for instance expects a population growth twice as fast as London (see Chap. 7).

Besides the fast growth of the population Australian cities have to deal with several other problems. Its car dependency (Kenworthy and Laube 1999) and high car use implies people do not exercise much hence it causes health problems such as obesity and diabetes. Additionally the distances travelled by car also influence the energy use and the net percentage of budget used by people living at distance from the city centre. In Sydney, the car travel from the Western suburbs therefore poses residents for a dilemma of keep on driving until it is hardly affordable anymore or move closer to the city centre where the petrol bill will be lower, but the affordability of the housing becomes a problem. This divide between western Sydney and the inner west, eastern suburbs and the northern beaches is evident, as mentioned before. Because car dependency is high, people also spend less time in green spaces, also influencing their health negatively (see Chap. 8). Moreover, in new neighbourhoods the amount of green space is low and not easy accessible. As a result of the increasing urbanisation the loss of biodiversity is significant (Aronson et al. 2014) and this is not recovered in the urban planning of new neighbourhoods.

Most of Australian major cities are located along the cost and are strongly influenced by different climate threats. In Sydney, sea level rise, storm surges and cyclones, (flash)flooding, prolonged heat and the heat island effect, and bushfires are seen as a serious threat (City of Sydney 2016).

Not only the energy use of high car use is high, the energy used for cooling/heating and electronic equipment in households is extreme, and, after the US, Canada, Gulf States and northern Scandinavian countries, one of the highest in the world (Wikepedia). The majority of energy is provided by coal-fired power plants. Only 2% is provided using rooftop solar (Origin 2015), so there is ample room for improvement (see also Chap. 9).

All these challenges are only to a limited level addressed in the process of urban design and planning.

1.3 Urban Planning and Design Education

The challenges as described above are not extensively addressed in the spatial design of new neighbourhoods and cities in Australia. One of the reasons may be that urban design is hardly taught at Australian Universities, and where it is it is

Table 1.1 Overview over urban design courses taught in Australia (Internet search)

University	Course title	Website
RMIT	Urban Design	https://www.rmit.edu.au/study-with-us/building/urban-design
University of Melbourne	Urban Design	https://msd.unimelb.edu.au/urban-design
Sydney University	Master of Urban Design	http://sydney.edu.au/architecture/study/urban-design.shtml
UNSW	Urban Development and Design	https://www.be.unsw.edu.au/degrees/postgraduate-coursework/master-of-urban-development-and-design
University of Adelaide	Master of Planning (Urban Design)	https://www.adelaide.edu.au/degree-finder/mpud_mplanud.html
University of Queensland	Urban Design	https://my.uq.edu.au/programs-courses/course.html?course_code=PLAN2003
James Cook University	Planning and Urban Design	https://www.jcu.edu.au/courses-and-study/international-courses/master-of-planning-and-urban-design
University of Western Australia	Master of Urban and Regional Planning	https://study.uwa.edu.au/courses/master-of-urban-and-regional-planning
Curtin University	Urban and Regional Planning	http://courses.curtin.edu.au/course_overview/undergraduate/urban-regional-planning
University of Southern Queensland	Urban and Regional Planning	https://www.usq.edu.au/study/degrees/engineering/urban-and-regional-planning
Griffith University	Urban and Environmental Planning	https://www2.griffith.edu.au/study/environment-planning-architecture/urban-environmental-planning
Macquarie University	Urban Planning – Environment	https://www.mq.edu.au/study/find-a-course/environment/urban-planning

often seen as an extension of an architecture study instead of being an own domain. Table 1.1 shows the seven urban design courses taught in Australia, and the five urban planning courses. Hence it is not very common to study the design of future urban environments, which could be seen as a decisive factor why the abovementioned challenges are not easily integrated in the process of urban development.

With such few urban design/urbanism education opportunities, certainly if this is compared with architecture offerings, it is clear the spatial design of cities is not much embedded in the cultural DNA of Australia. The results of 'this lack of attention' can be witnessed on a daily basis when driving around Australian cities and towns.

1.4 New Ambitions

The perspectives sketched in this book form the frontiers of urban thinking in Australia. These thought leaders have formulated cutting edge ambitions for current urban design practice.

The first element of good urban design is the attention for green space as a structuring element. Access to green spaces, or proximity, is an important factor for human wellbeing and health, and green spaces have many other benefits such as their influence on increasing resilience and pure ecological values. Therefore, green infrastructure, as Barbara Schaffer presents in Chap. 8, or an urban forest strategy, such as the City of Melbourne has developed (City of Melbourne 2012), and is followed by many other cities, is an important indicator of good urban design.

Secondly, urban density is a crucial factor in creating urbanity. Cities in which densities are lower, and sprawl is the dominating urban development, can hardly be seen as cities, rather as an extensive village. Currently, Australian cities belong to the cities globally with lowest densities (Chartingtransport 2016). In Sydney, but in several other Australian cities, policies, opinions and plans are developed to increase the density of urban areas. Not as a solution for everywhere, but in well chosen places, for instance close to railway or metro stations, or where the landscape quality is extraordinary. In the report 'Density done well' the link between high density and design quality of the public space is highlighted as a prerequisite for good urban design (Committee for Sydney 2016). A more conceptual approach is presented in Chap. 2, in which Craig Allchin illuminates the potential of Sydney's peninsula's to create higher densities, in a joint effort to establish better public transport and allow as many residents as possible to enjoy the beauty of their city.

Understanding the city as an organism, that is changing its shapes over time supports a view of the city to be able to adapt to new circumstances and requirements. No city is a fixed city, no matter what designers want you to believe, and thinking in end-use or –image id neglecting urban practice. Peter Bishop describes this in Chap. 4 as temporary urbanism (Bishop and Williams, 2012), and shows how small interventions and guerrilla methods can enhance the quality of the city. Temporarily, but that is what the city is. Under threat this is even more essential. Climate change is a reality and this means that the city should increase its resilience, as Hugh Gardner and Georgia Vitale demonstrate in Chap. 7.

The flipside of adapting to climate change is mitigation. The way we provide ourselves with energy determines the amount of carbon in the atmosphere hence temperature rising on earth. We can mitigate climate change by transitioning to renewable energy sources, reducing energy demand and reuse waste energy in the built environment, as Andy van den Dobbelsteen proposes in Chap. 9. Ultimately urban regions then could become zero-fossil, or, in a step towards becoming full zero-fossil, zero carbon (Roggema 2017; Roggema et al. forthcoming). Automotive, electric cars can contribute to this energy revolution, provided the electricity for the cars is generated from renewable sources, and not from coal fired power plants. Autonomous cars will become part of our city views sooner rather than later, as Jay Stricker demonstrates in Chap. 10. Not only will technology develop very quickly, the financial benefits for people (lower insurance, lower fuel costs), will undoubtedly push the sales and usage of these vehicles. Probably the regulatory basis is the single one aspect, which could slow down a fast implementation.

Apart from becoming more sustainable and resilient in the 'environmental' sense of the word, communities themselves can gain resilience by increasing their local economies. In Chap. 6 Justine Kinch is highlighting several examples around the globe how communities locally can improve their urban environment by developing micro-economies.

1.5 The Task Ahead

Looking at these ambitions it is evident the Australian approach to urbanism has an interesting future ahead of it. Where current urbanism is often based on ill-related silo's of policy making, such as separate transport, water and green infrastructure policies, which on their turn are set apart from spatial planning and design, a future urbanism should integrate all aspects of the urban environment people live in (Fig. 1.3). The aspects touched on in this book require integration at every scale of planning, from the metropolitan region to the local square. The task Australian urbanism stands for is to include all these aspects in the planning and decision-making process.

The chapters in this book each reflect in their own way and from different perspectives the required new approach to making cities in the Australian context. Most of current practice is baked in cultural convictions and traditions. However, this does not imply it is impossible to find new ways to enhance the urban quality of the cities Australians live in. Be this book an inspiration and an appeal to better urbanism at the same time.

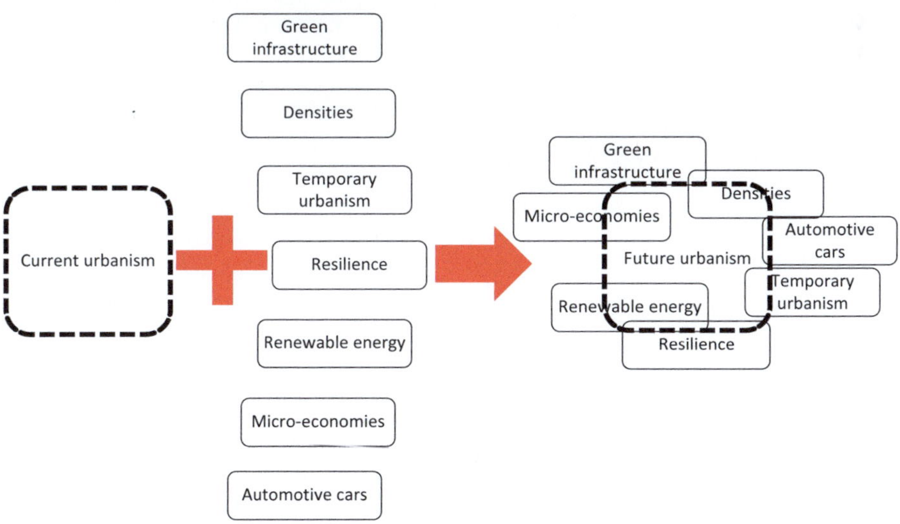

Fig. 1.3 From current practice towards an integrated approach for Australian urbanism

References

Albrechts L (2004) Strategic (spatial) planning re-examined. Environ Plann B 31:743–758
Alterman R (2001) National-level planning in democratic countries: a comparative perspective. In: Alterman R (ed) National-level planning in democratic countries: An international comparison of City and regional- policy making. Liverpool University Press, Liverpool
Aronson MFJ et al (2014) A global analysis of the impacts of urbanization on bird and plant diversity reveals key anthropogenic drivers. Proc R Soc B 281:20133330. https://doi.org/10.1098/rspb.2013.3330
Aubusson K (2017) Health tracker reveals Sydney suburbs with highest rates of obese and overweight children, inactive adults. Sydney Morning Herald. 2 May 2017. URL: http://www.smh.com.au/national/health/health-tracker-reveals-sydney-suburbs-with-highest-rates-of-obese-children-inactive-adults-20170427-gvtjys.html. Accessed 21 January 2018
BBR (Bundesamt für Bauwesen und Raumordnung – Federal Ministry for Regional Planning, Building and Urban Development) (1993) Law and practice of urban development in the Federal Republic of Germany. Bundesamt für Bauwesen und Raumordnung, Bonn
Bishop P, Williams L (2012) The temporary city. Routledge, London
Brenner N, Marcuse P, Mayer M (eds) (2012) Cities for people, not for profit: Critical urban theory and the right to the city. Routledge, London
Browne R (2016) Michael Lynch's plan to help close Sydney's social divide. Sydney Morning Herald. 16 May 2016. URL: http://www.smh.com.au/nsw/michael-lynchs-plan-to-help-close-sydneys-social-divide-20160513-gouuqc.html. Accessed 21 January 2018
Bunker R (2009) Situating Australian Metropolitan planning. Int Plan Stud 14(3):233–252
Chartingtransport (2016) Comparing the densities of Australian, European, Canadian, and New Zealand cities. URL: https://chartingtransport.com/2015/11/26/comparing-the-densities-of-australian-and-european-cities/. Accessed 21 January 2018
City of Melbourne (2012) Urban forest strategy. Making a great city greener 2012–2032. City of Melbourne, Melbourne
City of Sydney (2016) Resilient Sydney, Preliminary Resilience Assessment 2016. City of Sydney, Sydney
Committee for Sydney (2016) Making great places – density done well. Committee for Sydney. Sydney. URL: http://www.sydney.org.au/wp-content/uploads/2015/10/CfS-Discussion-Paper-Making-Great-Places-Density-Done-Well.pdf. Accessed 13 March 2018
Evers D, Ben-Zadok E, Faludi A (2000) The Netherlands and Florida: two growth management strategies. Int Plan Stud 5(1):7–23
Forster C (2006) The challenge of change: Australian cities and urban planning in the new millennium. Geogr Res 44(2):173–182
Freestone R (2007) A history of planning. In: Thompson S (ed) Planning in Australia: an overview of urban and regional planning. Cambridge University Press, Cambridge
Fuerst D (2003) Landes- und Regionalplanung Teil1: Verwaltungswissenschaftliche Grundlagen. Institut für Landesplanung und Raumforschung. Hannover. URL: http://www.laum.uni-hannover.de/ilr/lehre/skript_fu/lrp1.pdf. Accessed 20 January 2018
Gleeson B (2000) New challenges, new agendas for Australia's cities. Int Plan Stud 5(3):269–271
Gleeson B, Low N (2000) Australian urban planning: New challenges, new agendas. Allen and Unwin, Sydney
Greater London Authority (2016) The London plan. Greater London Authority, London
Hajer MA, Zonneveld W (2000) Spatial planning in the network society -rethinking the principles of planning in the Netherlands. Eur Plan Stud 8(3):337–355
Hall P (1992) Urban and regional planning, 3rd edn. Routledge, London/New York
Hanssen GS (2012) Negotiating urban space: The challenge of political steering in market- and network-oriented urban planning. Scand Pol Stud 35(1):22–47. https://doi.org/10.1111/j.1467-9477.2011.00278.x

Hofstad H (2013) Planning models in Sweden and Norway: Nuancing the picture. Scand Pol Stud 36(3):270–292. https://doi.org/10.1111/1467-9477.12006
Holston J (1995) Spaces of insurgent citizenship. Plann Theory 13:35–52
Infrastructure Australia (2015) Population estimates and projections. Infrastructure Australia, Canberra
Irvine J, Wade M (2017) The five charts that prove there is a housing affordability crisis. Sydney Morning Herald. 17 April 2017. URL: http://www.smh.com.au/business/the-economy/the-five-charts-that-prove-there-is-a-housing-affordability-crisis-20170414-gvla09.html. Accessed 21 January 2018
Kayden J (2001) National land-use planning and regulation in the United States: understanding its fundamental importance. In: Alterman R (ed) National-level planning in democratic countries. Liverpool University Press, Liverpool
Kenworthy JR, Laube FB (1999) Patterns of automobile dependence in cities: an international overview of key physical and economic dimensions with some implications for urban policy. Transp Res A 33:691–723
Križnik B (2014) Local responses to market-driven urban development in global cities. TEORIJA IN PRAKSA let. 51, posebna številka, pp 221–240
Kunzmann K (2001) State planning: a German success story? Int Plan Stud 6(2):153–166
Martin P (2018) Sydney, Melbourne move up on list of most expensive Cities Sydney Morning Herald. 18 January 2018. URL: http://www.smh.com.au/business/the-economy/numbeo-finds-sydney-king-of-the-costs-compared-to-melbourne-and-growing-20180117-h0jtvj.html. Accessed 18 January 2018
McCrindle (2015) Australia's capital cities: growth, change & a future forecast. URL: http://www.mccrindle.com.au/the-mccrindle-blog/australias-capital-cities. Accessed 21 January 2108
Miraftab F (2009) Insurgent planning: Situating radical planning in the global south. Plan Theory 8(1):32–50
Miraftab F, Wills S (2005) Insurgency and spaces of active citizenship: the story of western cape anti-eviction campaign in South Africa. J Plan Educ Res 25(2):200–217
Miraftab F, Silver C, Beard VA (2008) Situating contested notions of decentralized planning in the global south. In: Beard V, Miraftab F, Silver C (eds) Planning and decentralization: contested spaces for public action in the global south. Routledge, New York, pp 1–18
Næss P (2007) Normative planning theory and sustainable development. Scand Housing Plann Res 11(3):145–167. https://doi.org/10.1080/02815739408730358
Nilsson J-E (2007) Sweden – the emergence of a national urban policy. In: Van der Berg L, Braun E, van der Meer J (eds) National Policy Responses to urban challenges in Europe. Ashgate, London
NUMBEO (2018) Cost of living index 2018. URL: https://www.numbeo.com/cost-of-living/rankings.jsp?title=2018. Accessed 20 January 2018
Orchard L (1999) Shifting visions in national urban and regional policy. Aust Plann 36:20–25
Origin (2015) Energy in Australia. URL: https://www.originenergy.com.au/blog/about-energy/energy-in-australia.html. Accessed 21 January 2018
Popper FJ (1988) Understanding American land use regulation since 1970: a revisionist interpretation. J Am Plann Assoc 54(3):291–301
Randolph B (2004) The changing Australian City: new patterns, new policies and new research needs. Urban Policy Res 22(4):481–493
Regional Plan Association (2016) Charting a new course. A vision for a successful region. RPA, New York
Roberts P, Lloyd G (1999) Institutional aspects of regional planning, management, and development: models and lessons from the English experience. Environ Plann B Plann Des 26:517–531
Roggema R (2017) Sydney could be closer to being a zero-emissions city than you think. The Conversation. Published 15 December 2017. https://theconversation.com/sydneys-closer-to-being-a-zero-carbon-city-than-you-think-85976

Roggema R and van den Dobbelsteen A (forthcoming) Towards zero-carbon metropolitan regions: the example of Sydney. SASBE (under review)
Roy A (2005) Urban informality. Toward an epistemology of planning. J Am Plann Assoc 71(2):147–158
Saulwick, J. (2016) How Sydney's planners are using the 'Latte Line' to try and reshape the city. Sydney Morning Herald. 17 December 2016. URL: http://www.smh.com.au/nsw/how-sydneys-planners-are-using-the-latte-line-to-try-and-reshape-the-city-20161215-gtcfg5.html. Accessed 21 January 2018
Schmidt S, Buehler R (2007) The planning process in the US and Germany: a comparative analysis. Int Plann Stud 12(1):55–75
The Economist (2017) The global liveability report. The Economist Intelligence Unit, London. URL: https://www.eiu.com/public/topical_report.aspx?campaignid=Liveability17. Accessed 20 January 2018
Ting I, Bagshaw E (2016) HSC results 2016: Sydney divided by education 'latte line'. Sydney Morning Herald. 23 December 2106. URL: http://www.smh.com.au/national/education/hsc-results-2016-sydney-divided-by-education-latte-line-20161220-gtfhpf.html. Accessed 21 January 2018
Toon J, Falk J (2003) Sydney: planning or politics – town planning for the Sydney region since 1945. Planning Research Centre, Sydney University, Sydney
Troy L, van den Nouwelant R (2015) Sydney housing affordability index. UNSW, City Futures Research Centre. URL: https://cityfutures.be.unsw.edu.au/cityviz/affordability-index/. Accessed 21 January 2017
Van der Valk A (2002) The Dutch planning experience. Landsc Urban Plan 58:201–210
Wade M (2016) Census 2016: Sydney, a city divided by a yawning lifestyle and income gap. Sydney Morning Herald. 28 June 2016. URL: http://www.smh.com.au/business/the-economy/census-2016-sydney-a-city-divided-20170627-gwznsb.html. Accessed 21 January 2018
Wannop U (1995) The regional imperative: regional planning and governance in Britain, Europe and the United States. Jessica Kingsley, London
Wikepedia. List of countries by energy consumption per capita. URL: https://en.wikipedia.org/wiki/List_of_countries_by_energy_consumption_per_capita. Accessed 21 January 2018
Wong C, Ravetz J, Turner J (2000) The United Kingdom spatial planning framework, A discussion. The Royal Town Planning Institute, London. URL: http://www.rtpi.org.uk/media/6330/The-United-Kingdom-Spatial-Planning-Framework-A-Discussion.pdf. Accessed 22 January 2018

Chapter 2
The Future of Urban Planning in Australia: A Big Deal

Craig Allchin

> *Make no little plans; they have not magic to stir men's blood, and probably in themselves will not be realised. Make big plans, aim high in hope and work, remembering that a noble, logical diagram, once recorded, will never die, but long after we are gone will be a living thing, asserting itself with ever-growing insistency…Let your watchword be order and your beacon beauty (Reardon 1992).*
>
> Daniel Burnham, 1907 (Moore 1921)

Abstract Sydney's population is set to grow by over three million people in the next 40 years, going from around 4.7 million to 8 million (Greater Sydney Commission 2017a, b). One of the biggest challenges facing Sydney is managing this growth.

2.1 Introduction

Sydney's population is set to grow by over three million people in the next 40 years, going from around 4.7 million to 8 million (Greater Sydney Commission 2017a, b). One of the biggest challenges facing Sydney is managing this growth.

2.2 A New Plan

The Greater Sydney Commission (https://www.greater.sydney) has launched a new regional plan to manage this growth, which proposes a major reconfiguration of Sydney. Titled "Our Greater Sydney 2056 – A Metropolis Of Three

C. Allchin (✉)
Faculty of Design Architecture Building, University of Technology Sydney, Broadway, NSW, Australia
e-mail: craig.allchin@sixdegrees.com.au

R. Roggema (ed.), *Contemporary Urban Design Thinking*, Contemporary Urban Design Thinking, https://doi.org/10.1007/978-3-319-91950-8_2

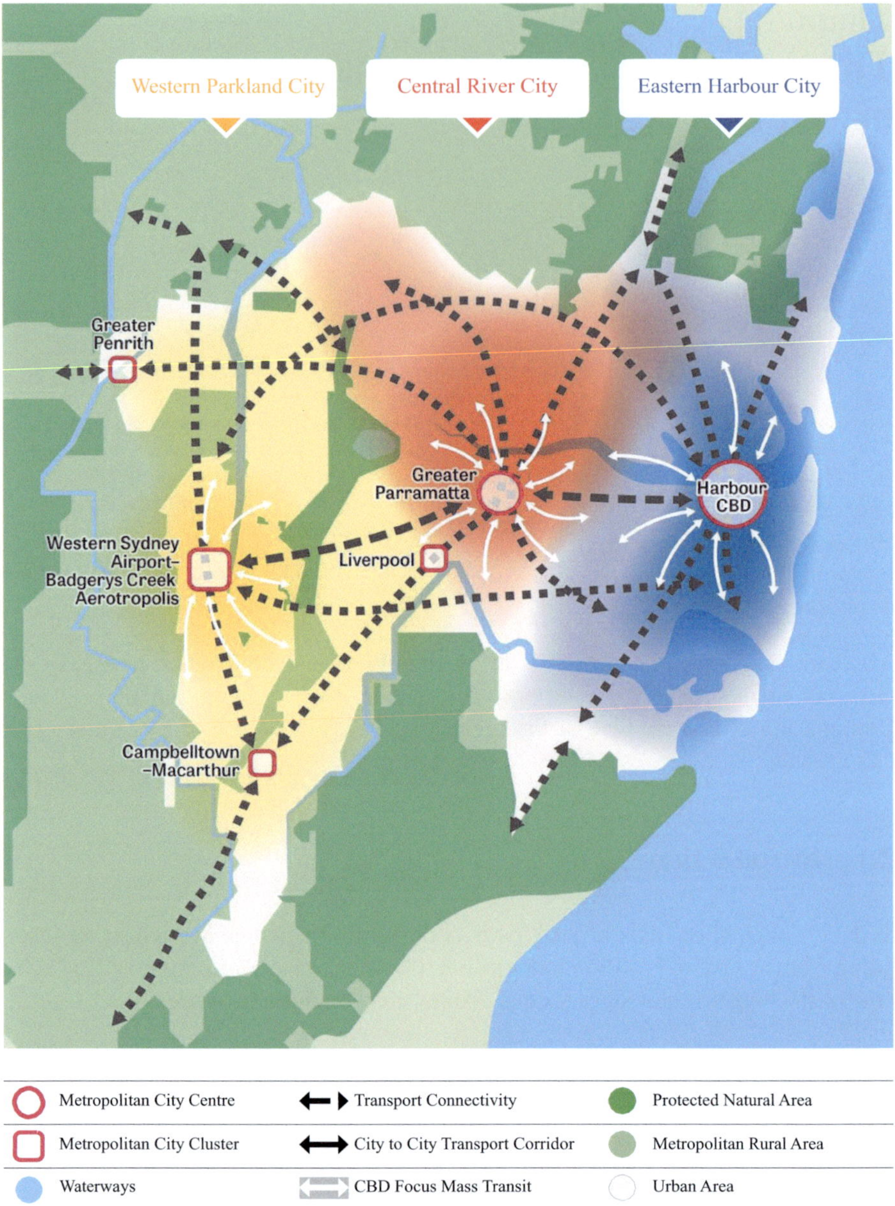

Fig. 2.1 Sydney's metropolis: three cities (Greater Sydney Commission 2017a)

Cities – connecting people" (Greater Sydney Commission 2017a) the plan proposes a spatial metropolitan concept of three distinct cities: an eastern harbour city, a central river city and a western parkland city (Fig. 2.1). The first two cities- the eastern and central cities are based around the original two settlements of Sydney, so it makes sense to continue to build around them, and create better access to them. The third city is a major new proposition, and raises several very big questions. The plan shifts the focus and growth to the west, with "almost half of the population living west of Parramatta by 2056 (Greater Sydney Commission 2017a, p. 6). Western Sydney is often 5–10° hotter than the east, has half the annual rainfall of eastern Sydney and is a long way from the signature Sydney landscapes of the harbour and rivers. However, from an (urban) development perspective, there is plenty of available, undeveloped land.

The new regional plan appears bold and visionary, proposing a fundamental change to the structure of Sydney by creating a new third city. But it must be examined whether it will create the best possible Sydney.

2.3 The Best Possible Sydney

High-density living is not often associated with proximity to spectacular nature. However two of the areas with the highest densities in Australia, Potts Point and Pyrmont, take advantage of their peninsula location to create some of the most desirable, and beautiful high-density mixed-use areas in the world. The built form has jostled for position over time to provide harbour views to many of the residences, whilst retaining good access to the main street of shops, cafes, services and transport that run along the ridges. In the case of Potts Point, taller towers have emerged more recently high up on the ridge (Fig. 2.2), whilst in Pyrmont larger

Fig. 2.2 Potts point. (picture: Craig Allchin)

Fig. 2.3 The Ultimo-Pyrmont plan aerial view. (source: Architectus)

buildings have been developed this century close to the water's edge (Fig. 2.3), on what became redundant industrial land before the 1990s. Other peninsulas close to the Sydney city centre, such as Kirribilli and Darling Point have also developed great density, whilst retaining their own characteristics.

Areas between Sydney and Parramatta, such as Five Dock, Concord, Rydalmere, Ermington and even Putney (Fig. 2.4) are similar in many ways to Potts Point and Pyrmont, with views of the water, access to ferries, and the potential for active, urban main streets along their ridges. If developed at high density these areas could provide to more people Sydney's unique lifestyle that combines the excitement and vibrancy of a major global city with the wonders and beauty of big nature.

When it comes to long term planning for the city, boldness is required. It must be explored whether some of the best parts of Sydney can be intensified to maximise their potential. A new regional plan should therefore focus on connecting the two original cities of Sydney and Parramatta, and intensifying the peninsulas in between them, rather than creating a new, dispersed, greenfields "city" in the west. For several reasons people make the choice to live in higher density environments that are closer to culture, transport and work. Intensifying the peninsulas between Sydney and Parramatta and connecting them with a circle metro line (Fig. 2.5), is a way to manage Sydney's growth *and* capitalise on the unique natural environment of the city. It is in the contrast between the dense urbanity and the rich nature where the magic of these quintessentially Sydney places is made.

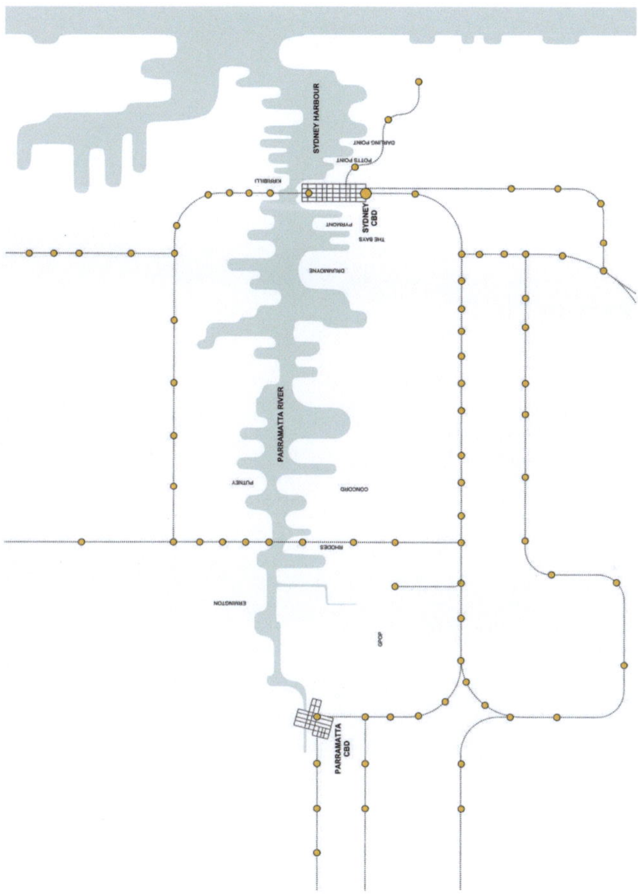

Fig. 2.4 Sydney diagrammatic, showing the peninsulas and connectivity. (Illustration: Craig Allchin)

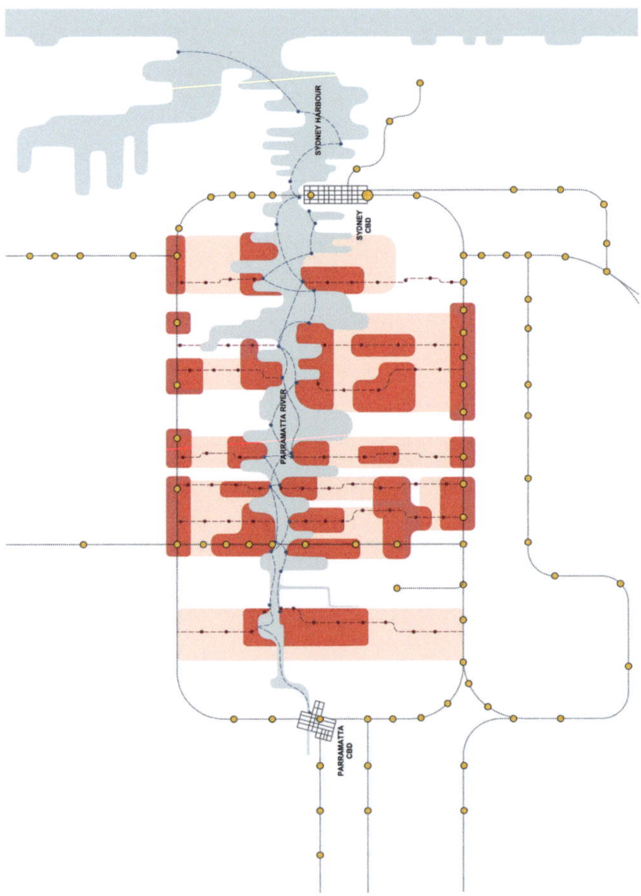

Fig. 2.5 Possible intensification of Sydney's urban system. (Illustration: Craig Allchin)

2.4 The Problem

The major problem with increasing densities on these peninsulas between Sydney and Parramatta is that most of these areas have already been settled, mainly with low-density housing. The majority of the redundant industrial land in this area has already been converted to residential uses, although there are two notable industrial areas remaining, the Bays Precinct and areas within the Greater Parramatta to Olympic Peninsula (Fig. 2.6), given the catchy acronym GPOP. Both these areas are currently being planned to transform into high density mixed use urban renewal precincts. Once these two areas are renewed, there will be very limited opportunity to increase the density around Sydney's central waterway.

In response to these difficulties the regional plan suggests development should move to where resistance is lowest, and find an easier way by shifting the focus westward to the third "Western Parklands City". Whilst some development should occur in the west, such as around the new airport at Badgerys Creek, this shouldn't distract from the main game, which is to create the best possible Sydney for the long term future. This requires the difficult task of transforming well located existing urban areas into new residential and mixed-use higher densities. However, to convert existing housing to higher density is much more complicated than converting redundant industrial land, or Greenfield land into low-density, easily developable neighbourhoods.

One of the main obstacles is political. A recent Sydney Morning Herald survey found that more than 66% of people think Sydney is full and development should be pushed outside of metropolitan Sydney. Only 23% supported more development in existing areas of the city (Nicholls 2017). This kind of very clear survey result makes it difficult for a government to ignore.

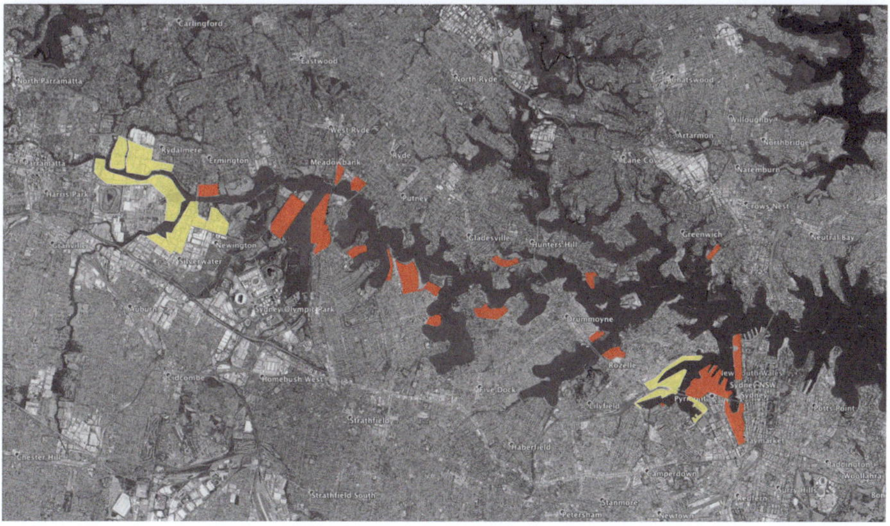

Fig. 2.6 Redundant industrial land in the Parramatta river precinct. (illustration: Craig Allchin)

Politically it would be easier to accommodate growth in new areas, where there are fewer stakeholders, such as, in this case, the Cumberland Plain where the majority of land is unsettled. Moreover, in the early twenty-first century, there is a definite and clear shift in lifestyle preferences away from the low-density distant suburbs, back to the city. People increasingly want to live, work and play close to the city centre. Ed Glaeser's Triumph of the City (Glaeser 2012) and Alan Ehrenhalt's The Great Inversion (Ehrenhalt 2012) are two of many books describing the twenty-first century's shifting preferences away from the suburbs and back to a more vibrant, engaged and urban lifestyle. Elizabeth Farrelly, one of Sydney's leading urban writers and thinkers, describes a new preference for "tighter cities and the walking lifestyle" (Farrelly 2017).

What does it say that almost 80% of people don't want any development near them yet at the same time most people say they want to live closer to the centre of the city? Could it be that the residents surveyed by the SMH who want no change are reflecting an anxiety with the process of change, and an uncertainty around the potential results? Do they also feel that the government may not protect their interests when developers come knocking at their door?

A new deal is required that puts more certainty around the processes of change, and enables residents to feel that their interests will be protected by the government if existing areas of the city are to be redeveloped.

The government must act in the long-term interests of the community, and the city as a whole. Sydney is the economic heart of Australia, responsible for 23% of GDP, and recently for 38% of GDP growth (Rawnsley 2017). It is also famous for its spectacular natural setting around Sydney Harbour and great ocean beaches. Should future Sydney continue to be based around these wonderful natural assets, or extend further away from them, into a new, hotter, western parkland city?

It is the responsibility of the built environment profession to work with the Government to paint a picture of the best possible Sydney and then explore ways to achieve it, through a long term regional plan, regardless of the perceived political risks. Leadership in difficult times is increasingly rare, but also increasingly important.

2.5 The Kind of City People Want

A successful regional plan has to understand the history of the city and address the big challenges of the future. It must also address the desired development of the city. In a way, it needs to capture the *Zeitgeist* and provide a flexible framework for the future of the city.

In 1948 the NSW government prepared the first long term regional plan for Sydney (County of Cumberland 1948). Titled "The County of Cumberland Plan" (CCP), it aimed to control the organic spread of the city by placing a green belt around it (Fig. 2.7). It was based on a similar plan in London implemented after the Second World War to manage the growth of the UK capital. The CCP plan worked

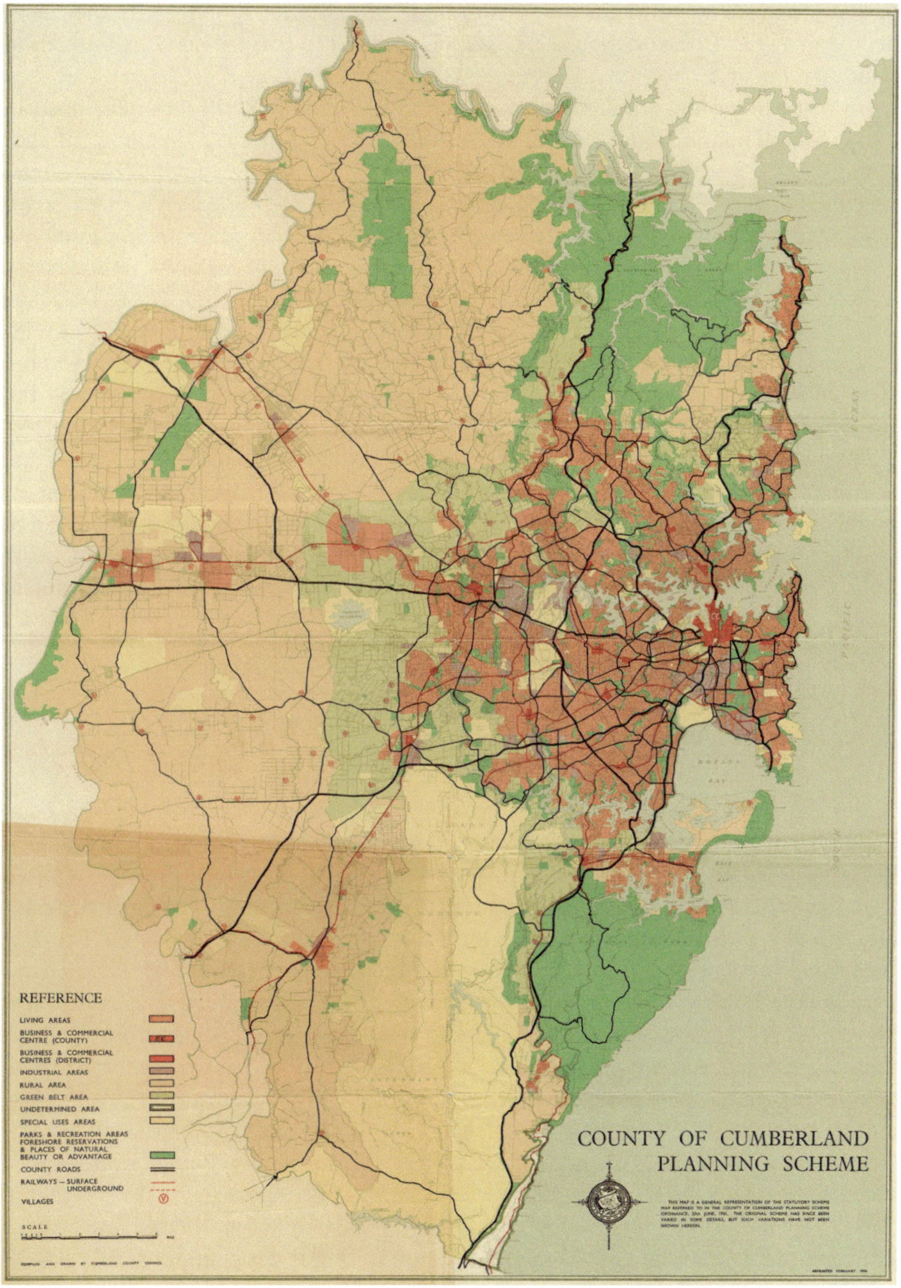

Fig. 2.7 County of Cumberland planning scheme. (Source: Collection of the State Library of New South Wales)

to some extent, however the suburbs jumped the green belt, driven by the unexpected growth of private motorcars and the demand to live the suburban dream in your own detached house.

In response to this, the Sydney Region Outline Plan (SROP) was developed in 1968 (State Planning Authority 1968), which was based on the Copenhagen finger plan of 1949. The SROP proposed corridors of development extending beyond the greenbelt southwest to Campbelltown, west to Penrith and northwest to Richmond, which had trains and highways running in parallel, like the structural bones of fingers, providing good access to the main city centre either by car or by public transport, with green space in between the fingers (Fig. 2.8).

The suburbs continued to grow along these fingers in the 1960s and 70s, providing wonderful lifestyles for families. This suited the average family structure at the time, which was different from now, with typically fathers working outside the home, and mothers within it. In fact, the suburban dream required one person (often: the mother) to be home most of the time because she created the social networks and bridged the physical gaps across the fences between each home, and between homes, schools and gathering places. The suburbs were designed for individual families, who wanted separation and their own space. The inevitable risk of isolation, which came with this physical structure, was overcome with a lot of social capital, generated mainly by women.

2.6 Urban Lives

Times have changed. In Sydney, the trend towards more urban lifestyles has been underway for the last two decades. Significant numbers of apartments have been built by converting redundant industrial land into high-density housing (Metropolitan Plan for Sydney 2036, Housing Chapter (2009)). This has mainly occurred in inner city suburbs, and on prime waterfront locations where port based industrial land has become redundant. Most of that land however has now been developed. The remaining industrial land is required to perform the task of servicing the city such as warehouses for food and commodities, places to repair cars and machines, and land for the utilities such as electrical substations. The question then becomes where can the people, including the young, singles, families and empty nesters, who want to live close to the city, find what they are after?

This brings us to the major conundrum faced by the Greater Sydney Commission when it prepared the regional plan. Where should new housing for the next three million Sydney-siders be developed: In existing eastern and northern areas, where most people want to live or, if previous planning trends are continued, in the west where access to undeveloped land is easier?

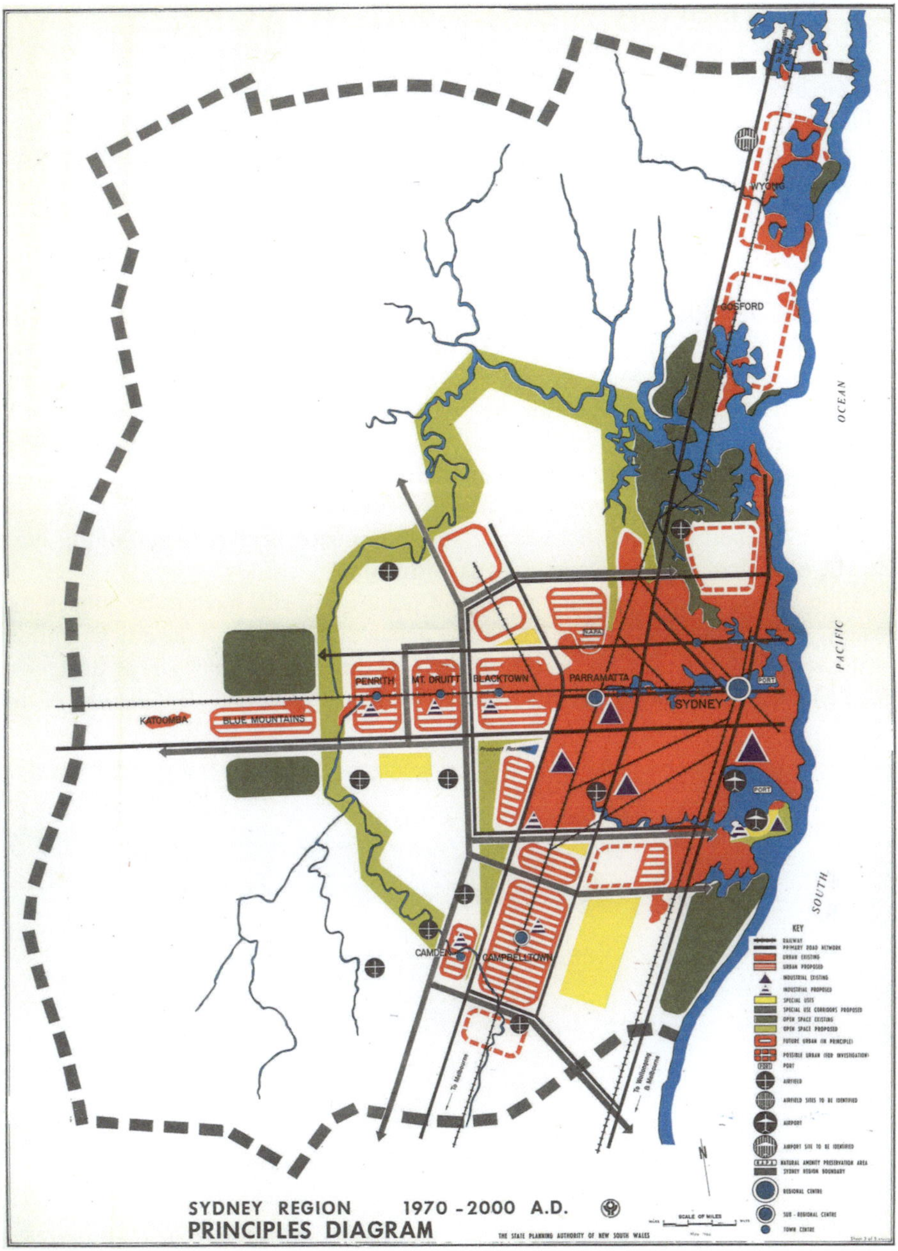

Fig. 2.8 Sydney Region Outline Plan. (source: State Planning Authority 1968)

2.7 The Third City

The Greater Sydney Commission has proposed what appears to be a clever, hybrid solution. The plan shifts most of the growth west into the uncontested land, limiting the political risk, and then calls it a 'city', to cater for the desire for more urban lives. It is possibly not as cynical as it first appears, as it might be feasible to create a great, high-density urban environments in the west (Fig. 2.9). But it will take a long time, it will be a very different form of city compared to the first and second cities, and it will be a very different type of nature that it is located within.

The Regional Plan states that it is "built on a vision where the people of Greater Sydney live within 30 min of their jobs, education and health facilities, services and great places" (Greater Sydney Commission, 2017a, b). It goes on to say that by 2056, it will have "rebalanced" the city, whereby almost half of the population will be residing west of Parramatta. In order to do this, the plan proposes that 392,000, or over half of the 725,000 homes required in the first 20 years, (the 2016–2036 strategic housing target), will be located west of Parramatta. Housing targets for 2036–2056 timeframe have not been set, but presumably well over half of this next tranche of homes will be located west of Parramatta.

Simultaneously, a significant number of homes will still have to be built in the eastern and central parts of the city, but the balance of the overall the city will move to the west. This seems a sensible strategy, particularly if you accept that most people in the established parts of Sydney don't want any change. But the role of the

Fig. 2.9 Western Sydney Employment Hub: looking towards the sites of the proposed Aerotropolis. (picture: Craig Allchin)

regional plan isn't just to keep the existing residents happy. It must act in the long-term interests of the citizens of today *and* of tomorrow, and ensure that the city will remain competitive, sustainable and equitable for everyone.

The Parkland City is a good title, but what sort of city will it be? There are large centres in western Sydney, but it is a creative leap to call the combination of Penrith, Liverpool and Campbelltown, (and the new Badgerys' Creek Airport) which are 15–20 km apart, a city (Fig. 2.10). It is, instead, a politically palatable way of accommodating the majority of Sydney's population growth in new greenfield areas, and in areas that will be much easier to develop than the existing well located suburbs of eastern and central Sydney.

Instead, the plan should address the problems with the current processes of urban renewal in the inner, more desirable areas. These cumbersome and haphazard processes have left the citizens of Sydney mistrustful of the development process and wanting no change in the existing areas. One of the particular problems is the Up-zoning of land around stations on major public transport routes, which is deeply flawed and has been unpopular.

2.8 Up-Zoning

Up-zoning is a process whereby the government changes the planning classification of land to allow denser development. Under a new higher or up-zoned classification, it becomes permissible to construct higher density built form, such as converting land with detached housing on it to town houses and medium rise apartments, or, in some cases, high rise apartment towers. The allowed type of up-zoning depends on the proximity to the station and the surrounding urban context. This rezoning process is underway around train stations in the Sydenham to Bankstown corridor of Sydney (Fig. 2.11), where an existing heavy rail train service is being converted to a more regular, higher capacity metro service (Fig. 2.12).

On maps, the NSW Government has shown the areas it is likely to up-zone (Fig. 2.13). However, doing so without enacting legislation, or putting in place checks and balances that might protect the existing residents. As a result developers are 'door-knocking' offering residents money for their homes, but the residents have no clue or certainty as to fair price. People don't know if they should take the money, wait for a higher offer, or perhaps not sell if they want to stay in their home. If the neighbours sell before them their home might become surrounded by high rise towers and lose value, so perhaps they should sell now. In this regulatory environment, residents are vulnerable, and don't support the process that makes them so.

Up-zoning has already occurred around stations on the Sydney Northwest metro line, in the Hills area (Fig. 2.14), and low-density residential land is being converted to higher density apartments. Some highly organized groups of residents have come together and sold their entire block to developers, sometimes making large profits up to two or more times the previous value of their homes, prior to the up-zoning, but this is relatively rare. More commonly criticism is heard of the governmental

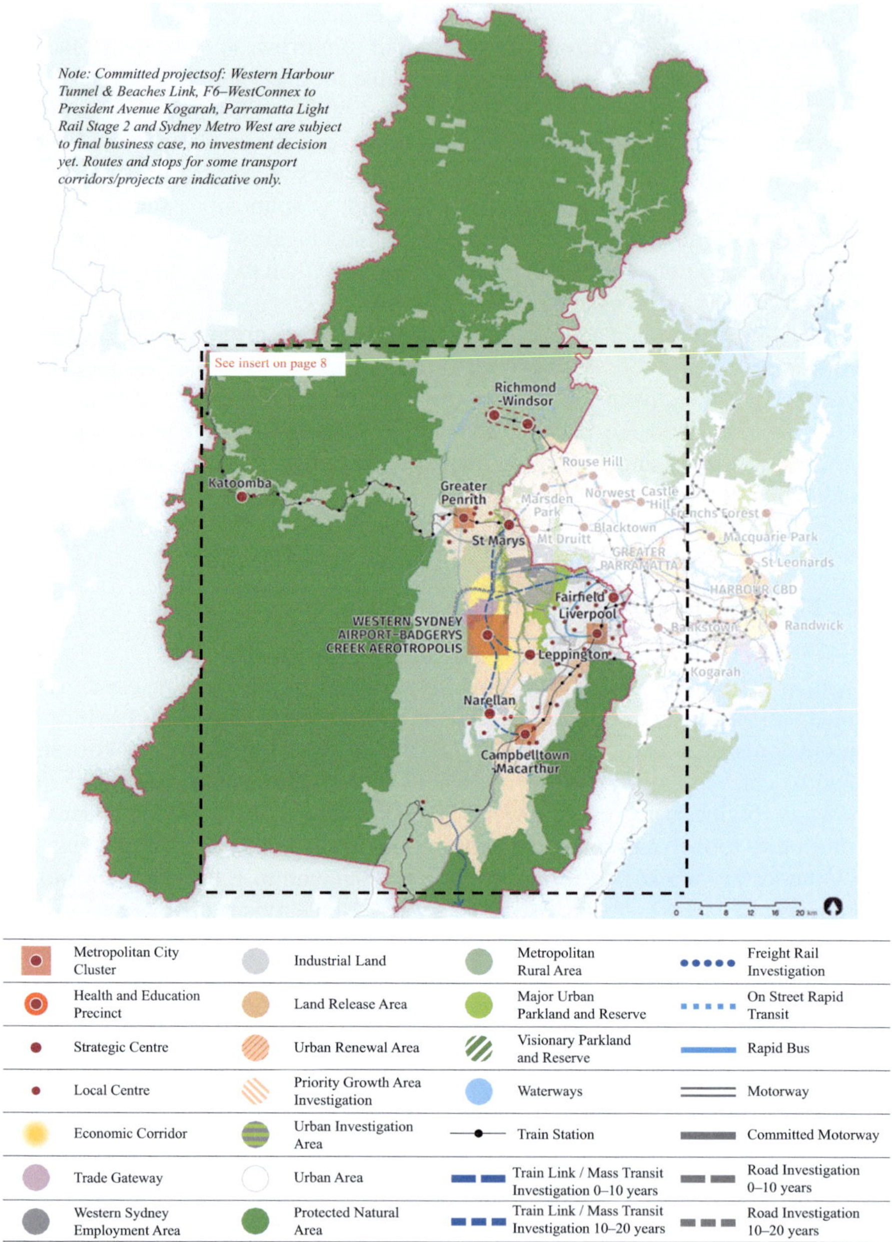

Fig. 2.10 Western Sydney District plan, draft. (source: Greater Sydney Commission 2017b)

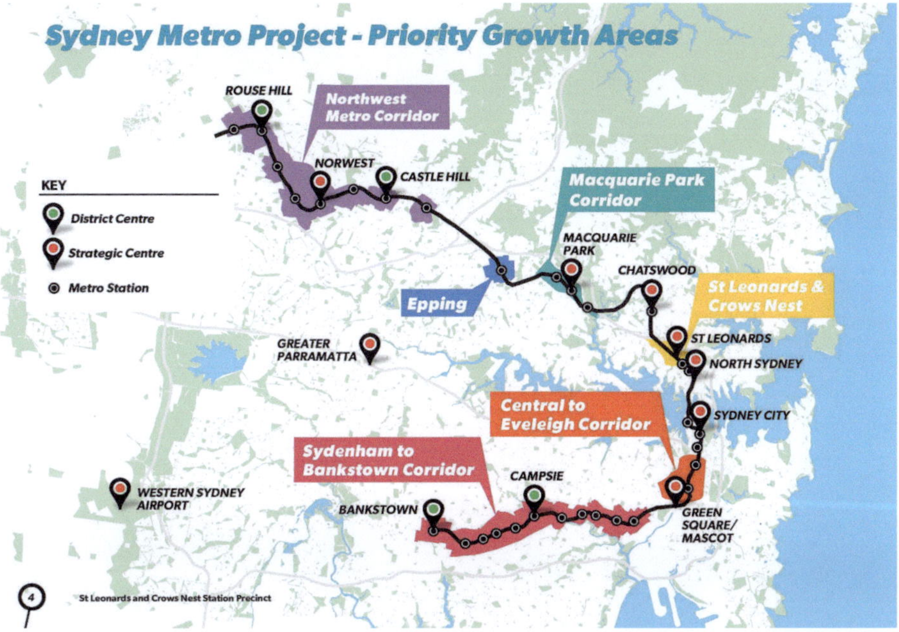

Fig. 2.11 Corridor Sydenham to Bankstown urban renewal corridor strategy. (Source: NSW Government Department of Planning and Environment)

up-zoning process and the uncertainty it creates for people. This could be an underlying argument why 66% of Sydneysiders don't want any new development in the city.

There is also the broader issue of land or funding not being set aside for additional schools, open space or other community services to cater for a significantly higher population that will be living in the up-zoned areas. The State Government has, in other locations, applied a developer levy on each new apartment, of perhaps $10,000, to cover some of the costs of service provision. This is not a very strategic approach. In fact, it is common place that once the land has been up-zoned without any public land set being set aside at all, the government has to pay full up-zoned market price for land to build schools, provide open space and other public benefits.

The process of up-zoning is clearly expensive, ad hoc and politically unpopular, unlikely to deliver optimum urban renewal in a highly contested city. Therefore it is understandable it results in people being anti-development, anti-developer, and suspicious of change. Rather than react to this by shifting the focus of development to another area in the west, far away from the city, options must be explored that deliver change in a better way.

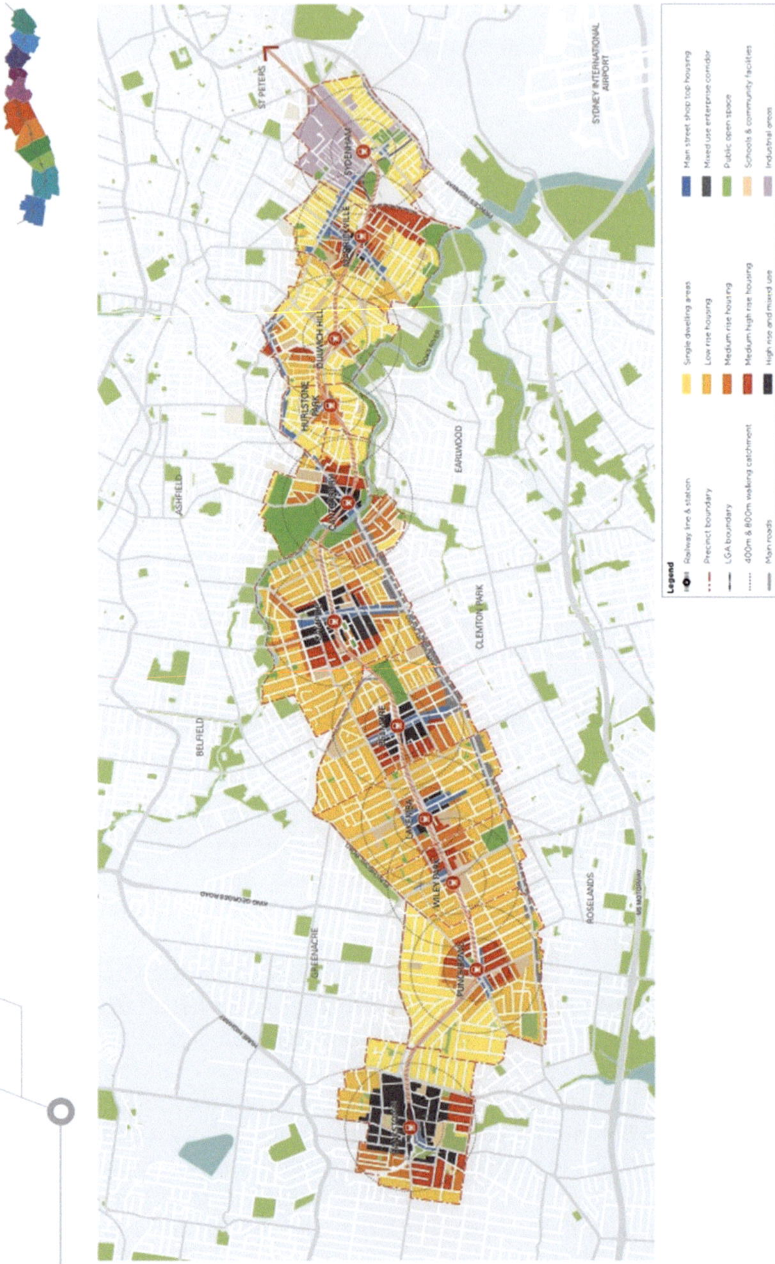

Fig. 2.12 Sydney Metro Pproject. (Source: NSW Department of Planning and Environment, St Leonards Crows Nest Interim Statement, 2017)

Fig. 2.13 Close-up Sydenham to Bankstown urban renewal corridor strategy. (Source: NSW Government Department of Planning and Environment)

2.9 A Better System: A New Deal

The people of Sydney should have a new deal put to them, which creates a more transparent, regulated and fair process to enable renewal of existing urban areas. The citizens of Sydney are well aware of the problems of housing affordability, particularly for the next generation. They are also very much aware that there is a shift away from the suburbs back to the city, and existing home-owners are acutely aware of potentially volatile property prices.

A new Sydney deal could be relatively simple. A development corporation, consisting of senior people from government, finance and development industries, should oversee the urban renewal of all key peninsulas between Parramatta and Sydney. Development corporations can be created under existing legislation (NSW Growth Centres Act, 1974). Significant areas of Pyrmont were renewed in the 1990s under the control of the City West Development Corporation (NSW State Archives and Records undated), which was incorporated into the Sydney Harbour Foreshore Authority in 1999 (Property NSW 2016), and the renewal continued. So a Development Corporation is not a new or an untested model for Sydney. However, the government owned most of the land that the Corporation controlled, so a new Peninsulas Development Corporation would firstly have to purchase the land to enable renewal at a significant scale. To do this, it could offer home-owners within designated areas double the current zoned value of their home (this to be confirmed

Fig. 2.14 Upzoning in the Sydney Metro Northwest area (upper), and upzoned area around the Showground Station Precinct. (Source Diagram by Craig Allchin (lower left), and Map NSW Department of Environment and Planning (lower right)

through a detailed financial modelling for every location specifically). This offer would be valid for between 5 and 10 years, indexed to CPI (Consumer Price Index), and could be funded through a partnership with Australia's Superannuation funds.

Typically, urban renewal in Sydney requires a tripling of the density to make it financially viable. On the peninsulas, densities should be increased between five- and tenfold in order to reach a similar density to Potts Point and Pyrmont. Therefore, this would generate sufficient funds to pay the home-owners an 'incentive-to-move' price, as well as to realise transport services, public open space and amenities, and appropriate profits to the superannuation and development industries.

A 5 or 10 year period would allow people to consider their options. If for example, children in a family were going to finish school in 6 years, they might choose to move, or if their work or lifestyle changed, this might influence the timing on their decision. During the 5 to 10 year period, homes owned by the development corporation could be rented out as affordable housing whilst structure plans are done. If people chose not to sell to the development corporation, then there would be an assessment process involving structure plan options, where some areas might be preserved, or interspersed with higher density development. Darling Point is a great example where large homes were preserved and interspersed with towers and medium rise apartment blocks, to accommodate the increased demand for well-located housing after the Second World War. There would inevitably be some compulsory purchases required to make each area work.

A bold new approach such as this would need to be tested in small areas to get the details right. A development corporation could be established as a test case on one of the Sydney peninsulas, ideally close to one of the new metro rail stations being planned along the West Metro route. Residents would be engaged to discuss the options of how they would see the future development. Digital imagery could be created of a possible future, informed by local and international case studies. A range of scales of development projects should be prescribed by the development corporation to ensure the ecosystem being created was varied in terms of housing types, scale and uses, much like the Potts Point and Pyrmont models.

This New Deal would create a vast, new and improved dream cityscape along our most valuable, liveable and delightful waterways.

2.10 Set by Infrastructure

Over time, and with considered transport and land use planning, Sydney could have an underground metro system, which approximately followed the original alignments of Parramatta Road and Victoria Road, encircling the Harbour and servicing all the peninsulas. The excellent but unfulfilled Christie rail plan of 2001 (Christie 2001) proposed Metro lines either side of the harbour although they weren't configured as a circle line (Fig. 2.15). If a Sydney Circle-line Metro existed, then light rail, buses and/ or perhaps autonomous public cars could connect the metro stations further up the peninsulas, with ferry terminals on the points running along vibrant main streets, creating a uniquely Sydney mesh of public transport options. That central transport network or mesh, would likely lead to a more dispersed pattern of employment that in turn could reduce some of the congestion problems of the city.

Many of the world's great cities have circle lines as a key organizing feature of their public transport system including Tokyo, Shanghai, Moscow, and London (Figs. 2.16, 2.17 and 2.18).

Central Paris had its big picture reshaping moment when Baron Haussman removed much of the medieval street pattern to create wide boulevards connecting major nodes (Fig. 2.19). This was done for a variety of reasons, including the need

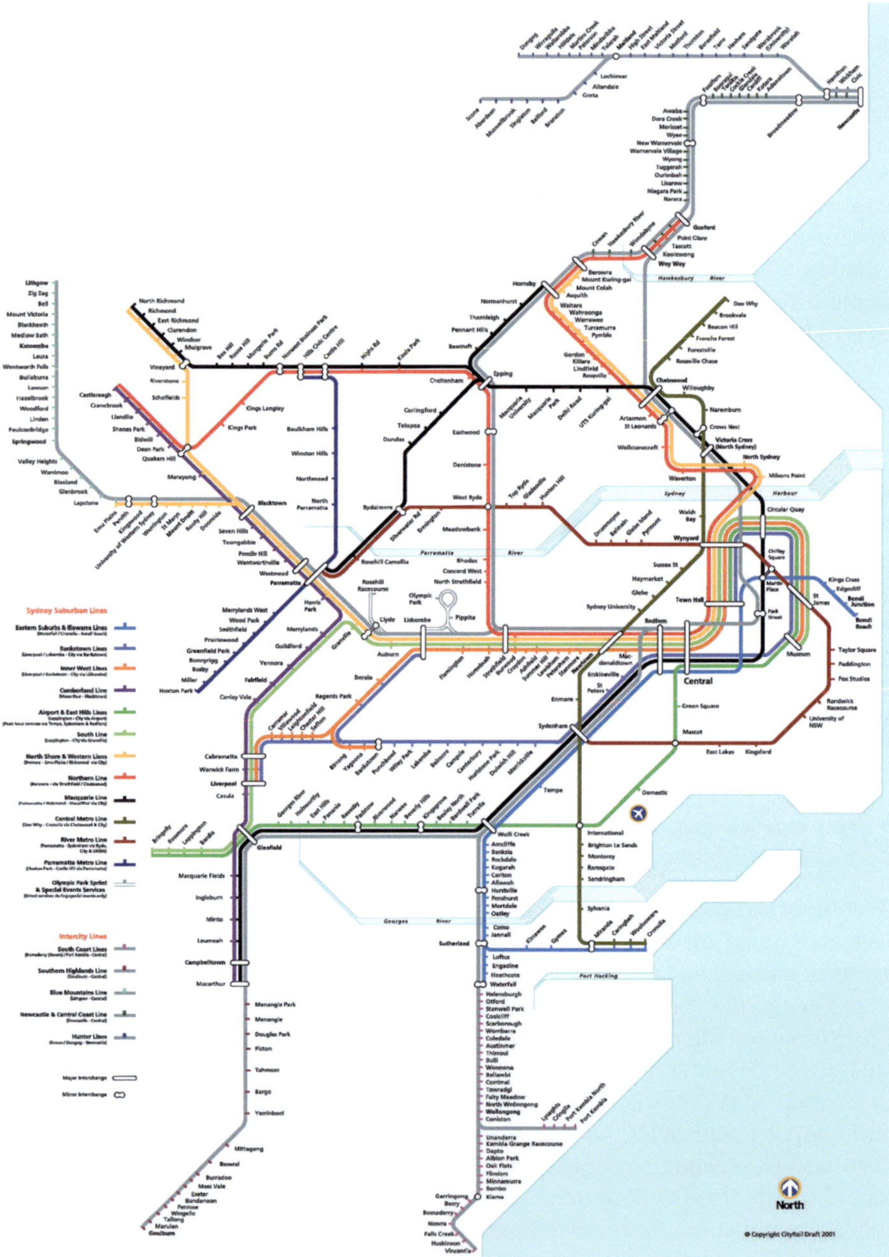

Fig. 2.15 Christie rail plan 2001. (Source: Christie 2001)

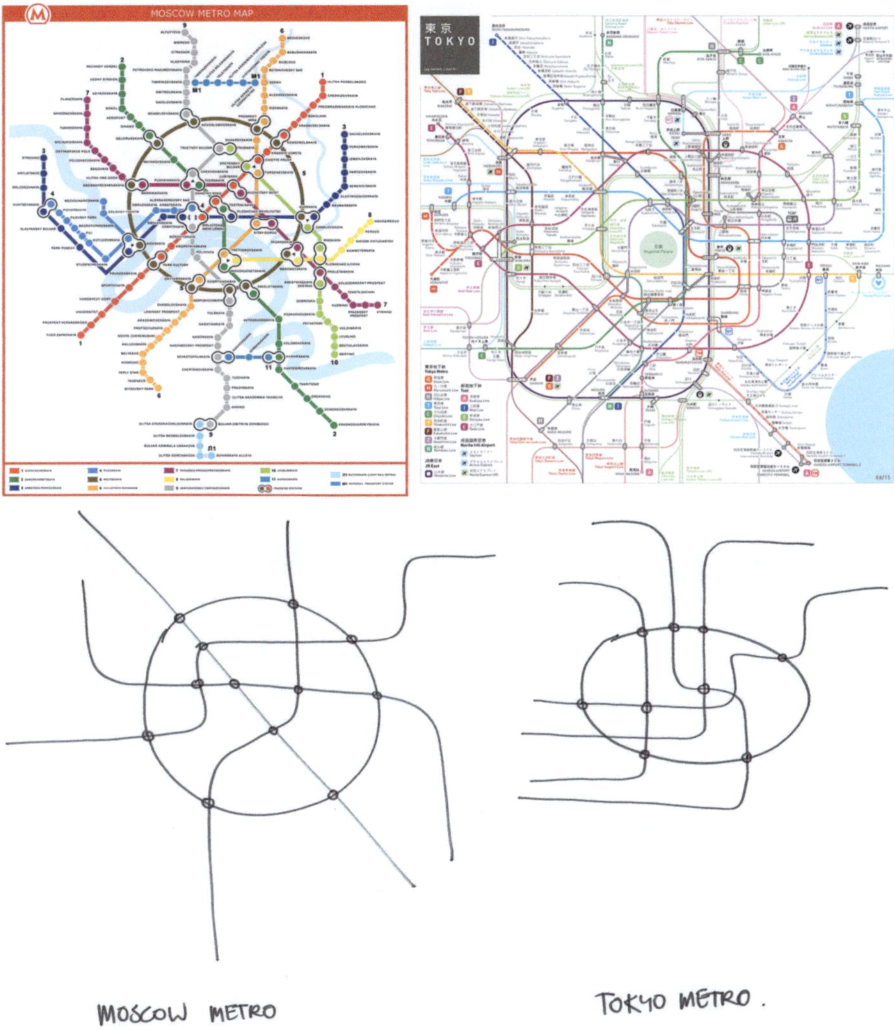

Fig. 2.16 Comparing metro-systems of Moscow and Tokyo. (illustration: Craig Allchin)

to achieve public order, but it created a lasting legacy of majestic boulevards that provide views to and connections between the key icons and places of the city. The grain and complexity of areas such as the Marais was retained, but an overall order imposed on the city.

In the same way, each of the Sydney peninsulas could be developed according to the local context and existing spaces and places, but with a new overarching structure that would provide a framework for growth and change around our greatest natural assets, The Harbour and The Parramatta River (Figs. 2.19 and 2.20).

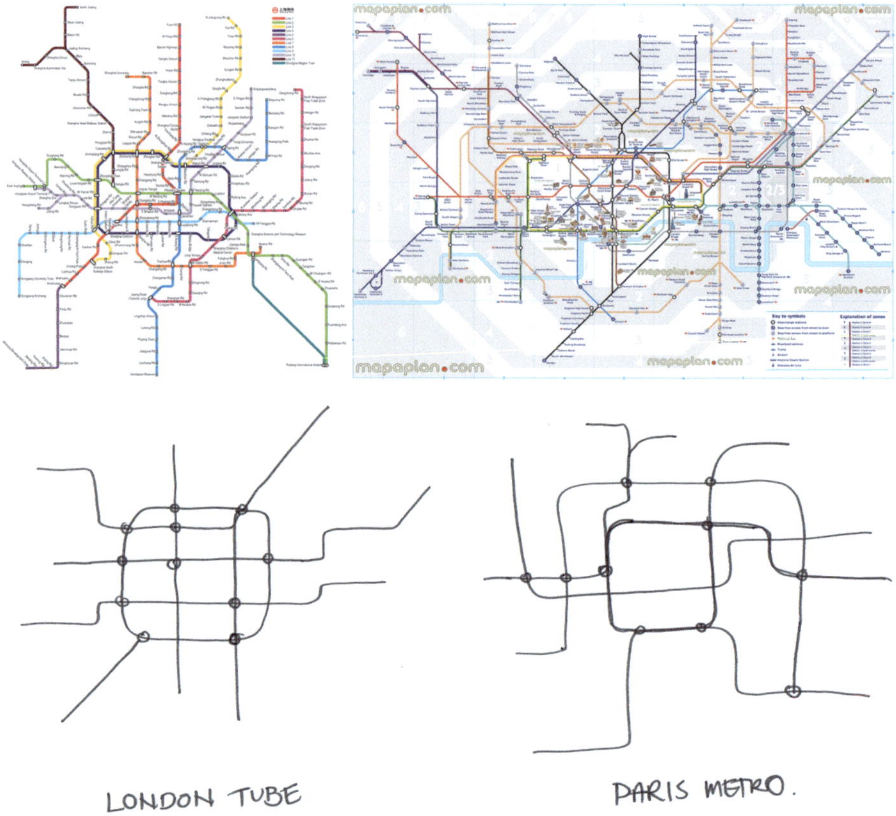

Fig. 2.17 Comparing metro-systems of Paris and London. (illustration: Craig Allchin)

2.11 Getting Over a Fear of Heights

Many urban planners in Australia and around the world praise the six storey Paris built form model and dismiss high rise nodes as less liveable, yet Potts Point, Pyrmont, Darling Point and Kirribilli have this interesting mixture of low, medium and tall built form (Fig. 2.21).

Central Paris, the area serviced by the Paris metro, which is approximately 10 × 10 km has a very different urban form to Sydney. Outside of this area Paris has sprawled and has many problems similar to other large global cities. But it is the central area that is best known and loved. In this area, most buildings are six stories high, built to the edge of the street blocks, many with shops and cafes on the ground floor. This form is the result of Baron Haussmans' reforms. Whilst this is a wonderful urban model and creates a consistent fabric of human-scale buildings, it is not the best model for Sydney, as it would require wholesale replacement of the built form. Sydney is closer to Tokyo than Paris, both physically and in terms of its urban structure.

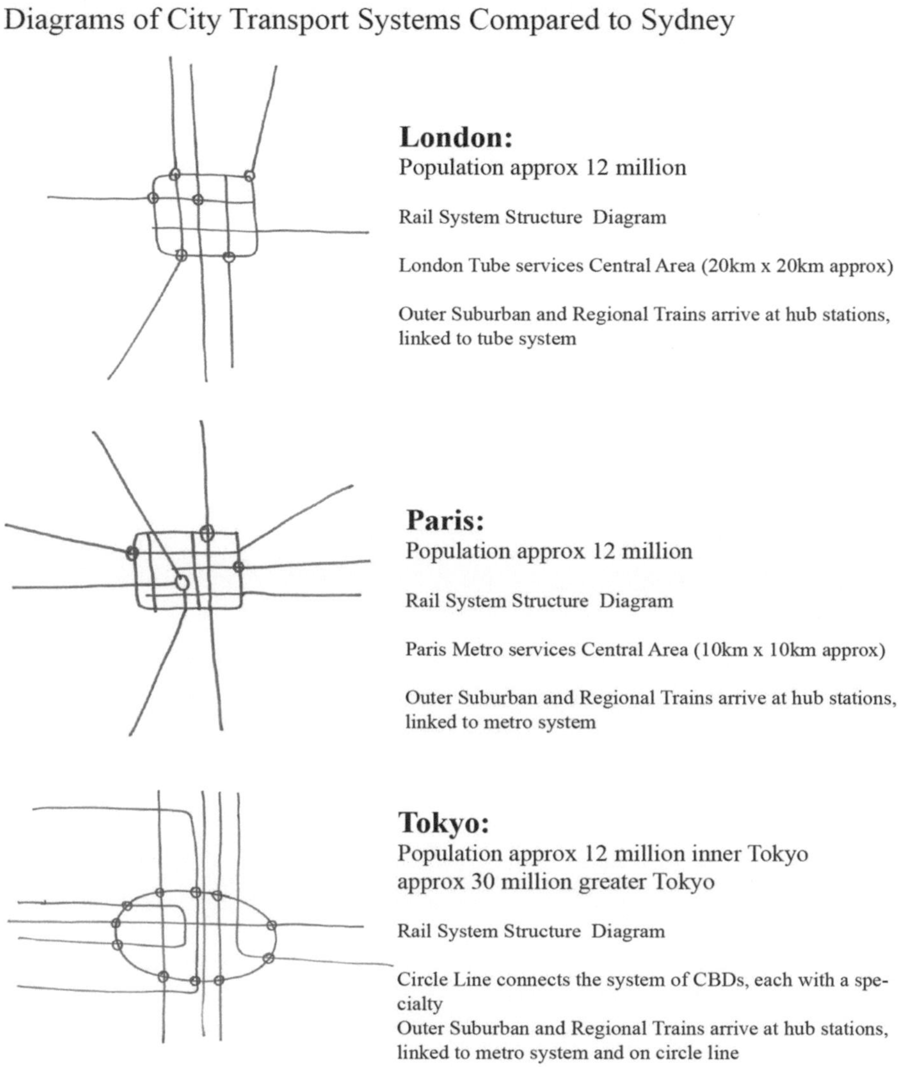

Fig. 2.18 Metro-systems of London, Paris and Tokyo. (Illustration: Craig Allchin)

Tokyo has a circle line, the Yamanote-line, which follow a roughly elliptical route of approximately 20 × 10 km. In the 1980s Tokyo had major congestion problems around its central station, so the government moved some of its offices to Shinjuku, a second centre on the Yamanote-line. Shinjuku, like many of the stations in Tokyo has very dense high-rise buildings around it yet within 800–1600 m there are small, detached houses on 100 m^2 blocks of land (Fig. 2.22).

Sydney has a similar built form pattern in some places, such as Chatswood, St Leonards, Parramatta, and Potts Point, but this has not been recognized yet as *the*

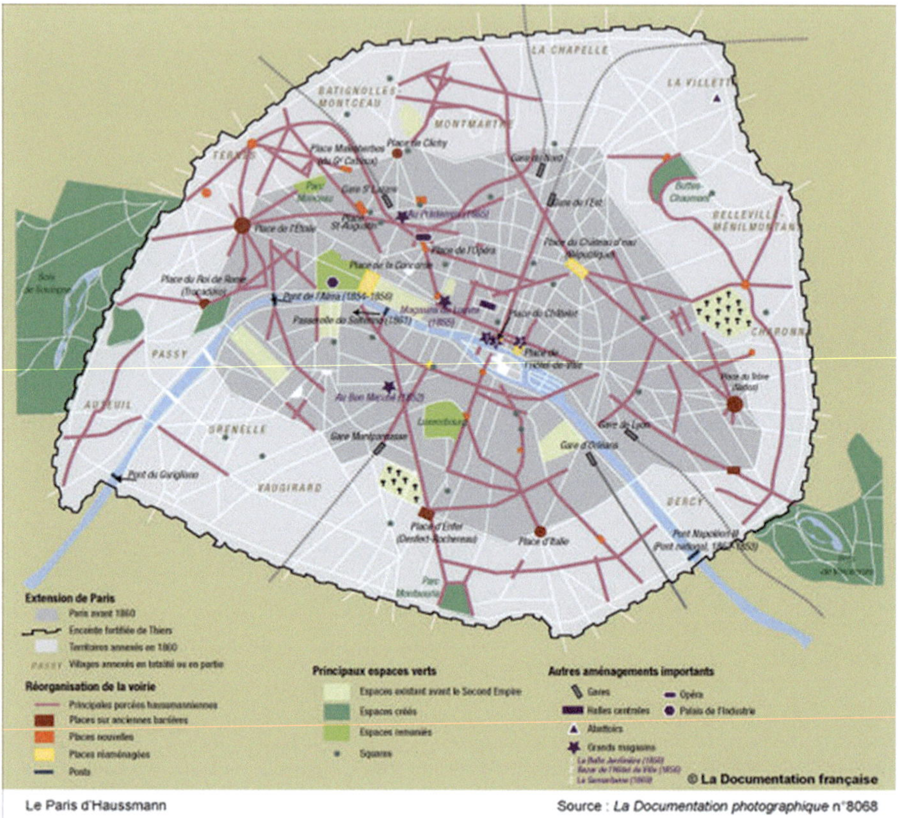

Fig. 2.19 Haussmans' Paris. (source: La Documentation Française)

ideal Sydney model (Fig. 2.23). Sydney's high density, mixed-use peninsulas could develop the exciting urban intensity and connectivity of Tokyo's circle line nodes. This would be further complemented and exaggerated, beyond what the Tokyo model has to offer, by the extra qualities of adjacent big nature of Sydney Harbour and the Parramatta River, which also add an additional layer of ferry transport to create a more connected and liveable city.

2.12 Conclusion

Sydney needs a big, bold vision if it is to become a better city for 8 million citizens in 2056, than it is today for 4.7 million. Development in Sydney is too contested and complex for the people to be pitted against the expert development industry, in a "let the market sort things out" way, with government watching from the sidelines.

Fig. 2.20 Low-density housing peninsula. (Source: Craig Allchin)

Fig. 2.21 Kirribilli. (Source: Craig Allchin)

Fig. 2.22 Shinjuku. (Illustration: Craig Allchin)

Fig. 2.23 From Parra to Sydney. (Illustration: Craig Allchin)

If the focus does shift to increasing the densities of the existing inner urban peninsula areas, previous models such as the up-zoning in the Sydenham to Bankstown corridor need to be replaced. Government should put a proper legislative framework in place that puts the people, the developers, and government on equal terms, and enables a serious, long-term engagement about the future. The legislation exists today to deliver this, and our superannuation industry is well placed to invest in creating and implementing new funding models.

The next generation of Sydneysiders, currently priced out of the housing market, should have the opportunity to live their Sydney dream. Being bold means suggesting they could have an even better life than previous generations, living in the best locations, utilising new technologies that can make Sydney more liveable, sustainable, productive and equitable?

Change in cities is slow, and the consequences of adopting the wrong plan can be serious and long term. Sydney is one of the world's great cities. As the guardians of the urban realm, the built environment professionals must act to advise, inspire and encourage the government to make bold plans that deliver on Sydney's potential.

As Daniel Burnham says: "let our watchword be order and our beacon beauty". A network of high density, vibrant peninsulas, connected across the spectacular central waterway of Sydney is the kind of "noble, logical diagram" that Burnham could have called for. Let's pursue this as our long-term plan, to make the best possible future a reality for Sydney.

References

Christie R (2001) Long-term strategic plan for rail: Overview report: Greater Sydney metropolitan region. Action for Public Transport New South Wales, Sydney

County of Cumberland (1948) County of Cumberland planning scheme. County of Cumberland

Ehrenhalt A (2012) The great inversion and the future of the American City. Knopf Doubleday Publishing Group, New York

Farrelly E (2017) Sydney town hall anti Westconnex event, October 2017. Authors recording

Glaeser E (2012) The triumph of the City. Penguin Random House, London/New York

Greater Sydney Commission (2017a) Our greater Sydney 2056 – A metropolis of three cities – connecting people. State of New South Wales. Sydney. URL: https://www.greater.sydney/draft-greater-sydney-region-plan. Accessed 2 December 2017

Greater Sydney Commission (2017b) Draft Western Sydney District Plan, connecting communities. State of New South Wales. Sydney. URL: https://gsc-public-1.s3.amazonaws.com/s3fs-public/web_gsc_draft_western_city_district_plan.pdf. Accessed 23 February 2018

Moore C (1921) Daniel H. Burnham, Architect, planner of cities, vol 2. Chapter XXV "Closing in 1911–1912". Houghton Mifflin, Boston, p 147

Nicholls S (2017) Sydneysiders in revolt over development as two-thirds declare the city is 'full'. Sydney Morning Herald. 9 October 2017. URL: http://www.smh.com.au/business/property/sydneysiders-in-revolt-over-development-as-two-thirds-declare-the-city-is-full-20171008-gywl1c.html. Accessed 10 October 2017

NSW Growth Centres (Development Corporations) Act 1974. http://www.ugdc.nsw.gov.au/about/legislation/ accessed February 2018

NSW State Archives and Records (undated) City West Development Corporation. URL: https://www.records.nsw.gov.au/agency/1091. Accessed 27 December 2017
Property NSW (2016) Sydney Harbour Foreshore Authority. URL: http://www.shfa.nsw.gov.au. Accessed 27 December 2017
Rawnsley T (2017) Economic performance of Australia's cities and regions 2016–2017. SGS Economics and Planning, Canberra/Hobart/Melbourne/Sydney
Reardon PT (1992) Burnham quote: well, it may be. URL: http://articles.chicagotribune.com/1992-01-01/news/9201010041_1_sentences-chicago-architects. Accessed 12 December 2017
State Planning Authority (1968) Sydney region outline plan 1970–2000 A.D.: a strategy for development. Government of New South Wales, Sydney

Chapter 3
How to Design Sydney's Third City?

Roderick Simpson and Rob Roggema

Abstract The development of a Draft Regional Plan for Greater Sydney (Greater Sydney Commission 2017) provides fertile ground for a reflection on current metropolitan planning practices and the relationship between metropolitan and local scales. The decision to establish the Western Sydney Airport (Australian Government 2017) has stimulated the idea of metropolitan Sydney being comprised of 'three cities' to improve equitable access to employment, education and cultural assets. The first city is characterised as the 'Eastern harbour city', the second is centred around Parramatta as the 'Central river city' and the future third city as a 'Western parkland city' (Fig. 3.1).

3.1 Introduction

The development of a Draft Regional Plan for Greater Sydney (Greater Sydney Commission 2017) provides fertile ground for a reflection on current metropolitan planning practices and the relationship between metropolitan and local scales. The decision to establish the Western Sydney Airport (Australian Government 2017) has stimulated the idea of metropolitan Sydney being comprised of 'three cities' to improve equitable access to employment, education and cultural assets. The first city is characterised as the 'Eastern harbour city', the second is centred around Parramatta as the 'Central river city' and the future third city as a 'Western parkland city' (Fig. 3.1).

There is no obvious connection between the more fine-grained and localised concerns of urban design and the metropolitan scale other than a desire and necessity to make the third city as liveable as possible, with urban design as part of a

R. Simpson (✉)
Greater Sydney Commission, Sydney, Australia

University of Technology Sydney, Ultimo, Australia
e-mail: rodericksimpson@sw.net.au

R. Roggema
Faculty of Design Architecture Building, University of Technology Sydney, Broadway, NSW, Australia

Cittaideale, Wageningen, The Netherlands
e-mail: rob@cittaideale.eu

R. Roggema (ed.), *Contemporary Urban Design Thinking*, Contemporary Urban Design Thinking, https://doi.org/10.1007/978-3-319-91950-8_3

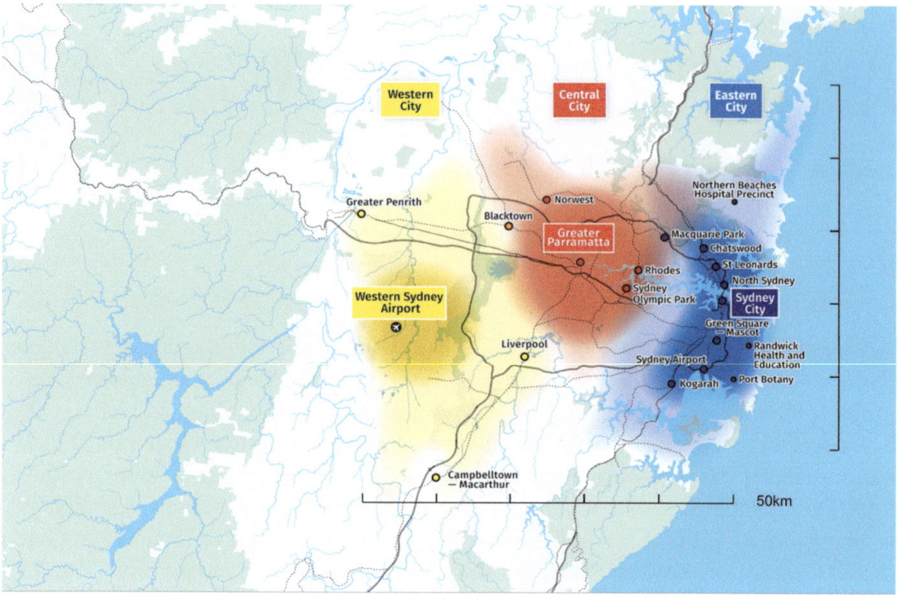

Fig. 3.1 The three cities in metropolitan Sydney. (source: Greater Sydney Commission 2017, scale added by authors)

linear planning and delivery process. Metropolitan strategies tend to focus of 'city-shaping infrastructure', principally major transport needs. However too narrow a focus on these large infrastructure projects risks overlooking three critical aspects:

Firstly, the way cities are built and rebuilt and what makes up the majority of investment and activity is not the major structuring infrastructure projects that fundamentally alter the economic geography of the city. While these make new areas competitive and able to be inhabited, most of the investment, the 'filling in', is realised on a place-by-place basis. How that is conceived, delivered and governed is what will determine the character and liveability in the longer term.

Secondly, in Sydney the peri-urban area is largely in private ownership but has a wide range of public values. This raises the challenge of both recognising and planning at a landscape scale larger than major infrastructure projects, and gives rise to the need for an overall concept for the landscape as 'green infrastructure', to serve the expanded city but also to preserve its intrinsic and ecological values. Clearly, the heterogeneous patchwork of landscape values, constraints and landowner expectations does not lend itself to a reductive McHargian assessment of land capability and even less to a rational land use framework that might result. Instead, a projective and aspirational concept for the landscape is needed, that recognises, respects and collaborates with the wide range of interests to develop such a shared vision. This aspiration is contained in the idea of the 'parkland city', recognising the landscape is subjected to human intent for many generations long before European occupation and that nature, particularly in an around urban areas cannot look after itself, is subject to human decisions and intention, and these intentions need to be made explicit. An aboriginal statement expresses the attitude best: 'if we care for county, country will care for us', in scope and recognition of the interdependencies relating to the concept of 'socio-ecological systems' (SES).

Thirdly, planning for major infrastructure is firmly grounded in conventional engineering practice that tend to mechanistically assume that the future is predictable, based on the continuation of current trends that allows forecasts of demand, patronage and capacity. This way of seeing the city may be contrasted with the actual history and conditions in the city, which are uncertain and chaotic. In neoliberal systems, where the contraction of public services and their accompanying transfer and transformation to fit with business models, certainty is essential and may be assured by the introduction of anti competitive measures to ensure the profitability of a particular project or infrastructure. In other words, the centralised technocratic controlling tendencies of modernism are essential to the operation of neoliberalism, which depends even more than government (which has greater capacity to respond to perturbations and uncertainty) on predictability, codification, limits to competition and disruptive innovation.

3.1.1 Tensions

These three aspects highlight a number of tensions inherent in planning and urban design: between top-down and bottom-up processes, between public and private interests, between the ideal and the existing, between the deadening desire for certainty, and the unsettling but fertile uncertainty of the ways the city actually operates, between the desire and need to 're-direct' the city and the momentum and inertia of the existing physical city and the path dependencies of institutional, bureaucratic, social and political arrangements.

Given that these tensions can never be 'resolved' because there is no definable end-state, it is important to recognise the limitations of, indeed the impossibility of achieving perfect hierarchical plans. This suggests that different modes of practice for urban design are required. Where there is tension, there is a professional opportunity! Instead of having an instrumental role in making and 'delivering' hierarchical plans, urban designers might be seen as intermediaries in the dialectic process of top-down and bottom-up processes where on-going governance is as important if not more important than the plan-as-product itself.

3.1.2 A Never Ending Experiment

Moreover, the city is not a complicated but predictable machine, and instead complex and self-organising, inherently unpredictable. Uncertainty has been characterised as a 'wicked problem' that is apparent in all social systems, urbanisation and city making being just one. Given that there can be no defined end state or 'stopping rule' to use Rittel and Webber's terminology (Rittel and Webber 1973), and only better or worse outcomes from any particular intervention in the city, the city can then be seen as a never ending series of experiments at varying scales. This in turn gives rise to the question of who has control or defines the experiments, and how are outcomes judged as better or worse.

These considerations lead to distinguishing the roles of 'top-down' large projects and government- the structuring physical and information framework on the one hand side, complemented by 'bottom-up' experimentation, innovation that is made possible by relative autonomy and localism, and an essential attribute of resilience (Walker and Salt 2006) as well as a sense of place (Friedmann 2010) on the other.

In this chapter the first question discussed, is what constitutes a 'good city'. In order to reveal this we investigate the first organically grown city and the functionalist second city respectively. Learning from these and by adding resilience as a framework for planning we formulate the conditions for designing the third city and redefining the field of operations for urban design.

3.2 A Good City Is a Resilient City

What constitutes the 'good' city can be considered at a number of scales. At the local scale urban design is usually associated with walkability, identity, authenticity and amenity, aspects that figure prominently at conferences, in literature, in manuals, guidelines and codes.

At a metropolitan scale aspects such as efficiency, relative equality, productivity, or the optimisation of infrastructure are emphasised. There is considerable path dependency in metropolitan plans themselves. A culture and professional practice that relies heavily on multinational consulting firms, league tables produced by academia and multination quangos, that go to underpin a continual jockeying for position and an underlying sense of insecurity - 'standing still is to fall back in the pack'. For 'global cities'[1], these documents often refer to the categorisation of alpha, beta and gamma global cities based on narrow criteria related to business networks and connectivity. These plans tend to have four domains: economy, housing, environment, and implementation.

This is where thinking about what constitutes the good city usually stops. However international competitiveness itself is really subordinate to the ultimate existential measure of the good city: the city that survives, persists and thrives, which may be characterised as the resilience of the city to both endogenous (demographic change, spatial inequality, housing affordability) and exogenous (climate change, artificial intelligence, globalisation). shocks and stresses.

Arguably it is the exogenous mega-trends: regional instability driven by the impacts of climate change (Department of Defence 2016), population growth, Artificial Intelligence (AI) and financial system crashes that may have the greatest potential impacts on the sustainability and resilience of the city. For developed countries like Australia the impact of natural disasters is likely to be far less than that of human induced impacts of the megatrends, outlined above, and yet, as has

[1] Sydney has 'slipped' from alpha+ to alpha since 2012 http://www.lboro.ac.uk/gawc/

always been the case issues close to home that are immediately apparent dominate public perceptions and discourse. The difference may be in the 'acuteness' of the impact- the mortality rate, but on any other measure these other 'chronic' stresses have a greater impact in relation to the full potential of the city (Wilkinson and Pickett 2011; Stiglitz 2015). The good city must therefore not only provide wellbeing for its citizens, it must also be a 'good global citizen' itself, which does not focus purely on competitiveness without 'giving back'. In order to do that it must have the capacity to be generous and to contribute to global wellbeing.

"Urban **resilience** is the capacity of individuals, communities, institutions, businesses, and systems within a city to survive, adapt, and grow no matter what kinds of chronic stresses and acute shocks they experience" (100Resilient Cities undated). This reframes what constitutes the 'good' city in terms of 'resilience': urban design needs to consider the formulation of governance also to achieve 'resilience'. This resembles the shift from 'urban design' to 'urbanism'.

Despite this, few cities have yet have adopted 'resilience' as the principal guiding framework to structure a strategy although almost all include disaster preparedness and planning for the impacts of climate change at a local level. This is understandable given the path dependencies in the way these strategies are prepared as outlined above and in the way they have to be translated into implementation that can fit with, and be delivered by, existing institutional structures on the one hand and the difficulty of formulating, advocating, adopting and communicating a global perspective on the other.

In short, most of the strategies attempt to do too much, are based on older models of 'predict and provide' and fail through overreach, but not enough in a global context.

We contend that replacing the skeletal remains of centralised modernist functionalism with an overarching 'resilience' framework will help address most if not all of the tensions outlined above.

3.3 General Characteristics and Attributes of Resilient Systems

It is not the aim of this paper to provide an overview of resilience thinking (Walker and Salt 2006; Holling 1973; Biggs et al. 2015; Resilient Cities 2012; Newman et al. 2009; ARUP 2014; UNISD undated) or even its application to urbanism in general (ARUP 2015; UNISDR undated; 100RC undated). Its aim is to look at how the scope of urban design practice might be repositioned if urban resilience is adopted as a key driver.

There have been attempts to transfer 'resilience' directly to urban design (Resilient Cities 2012; Feliciotti et al. 2016), but these do not give sufficient emphasis to the governance and learning aspects of resilience. It will suffice to look at the generally agreed attributes and processes that contribute to resilience. These are summarised in the Table 3.1.

Although these frameworks are not perfectly aligned, there is sufficient commonality. Attributes such as redundancy, diversity and modularity can be found in

Table 3.1 Attributes and processes related to resilience (ARUP 2015; 100RC undated; Biggs et al. 2015; Resilient City 2012; Walker and Salt 2006)

ARUP	100RC	Stockholm Resilience Centre (Biggs et al.)	ResilientCity	Walker+Salt
Accepting of uncertainty +Adaptive	Flexibility	Foster complex adaptive systems thinking	Capacity for Adaptation	Ecological Variability
Reflective	Reflectiveness	Encourage learning	Feedback Sensitivity	Tight Feedbacks
Resourceful efficient	Resourcefulness			
Robust (hardened)	Robustness			
Diverse		Maintain diversity	Diversity	Diversity + Innovation
	Redundancy	Maintain redundancy	Redundancy	
Integrated	Integration	Manage connectivity	Integration	
Inclusive	Inclusiveness	Broaden participation		Social Capital
		Promote polycentric governance systems	Modularity and Independence of System Components	Overlap in Governance + Modularity
		Manage slow variables and feedbacks	Environmental Responsiveness	Acknowledging Slow Variables

the physical fabric of the city, while processes, such as networked communication and learning (monitoring/slow feedbacks) and subsidiarity (polycentric governance) are at work in different parts of the city, at different scales and under different planning regimes (Fig. 3.2).

3.4 The 'Organic' First City

The idea of the 'organic' city, as a result of self-organisation, prompts images of medieval hill towns in Europe, the *casbahs* of the Maghreb, smaller agricultural towns in Asia and more recently the favelas of South-America and the slums of Africa. However, the apparent 'organic' quality they share masks very different underlying codes that have given rise to the forms.

In the case of Sydney, very shortly after European occupation, five sets of land tenure operated: the customary stewardship of the land by the Aboriginal population, land claimed for the Crown (separated into plots for various activities and the initial

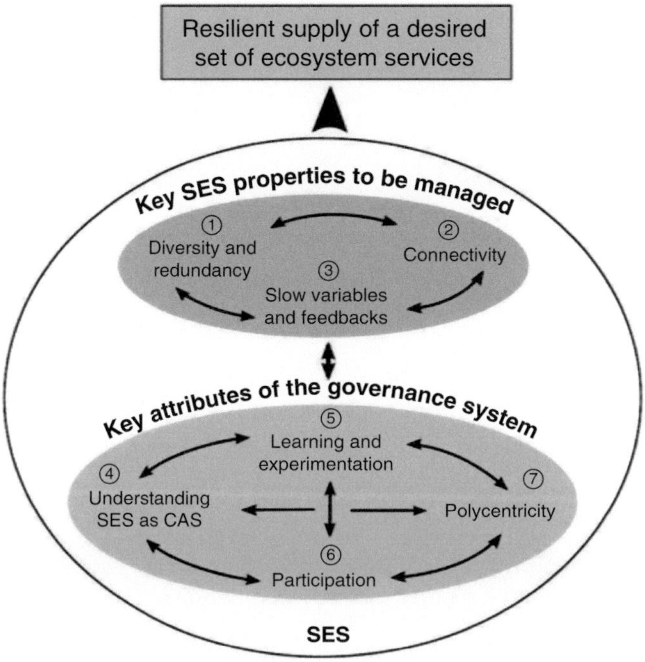

Fig. 3.2 Seven generic policy-relevant principles for enhancing resilience. (source: Biggs et al. 2015)

'public domain' of streets distinct from the state 'domains' and the "Sydney common"), land grants to commissioned officers, the informal settlement of the Rocks (1788+) and slightly later the first free land grants (1792) (NSW State Archives and Records undated). The process and motivation for subdivision is critical for the self-organisation of the city (Fig. 3.3).

Kings Cross and Potts Point in central Sydney was until recently[2] the densest urban area in Australia with about 15,000 residents and 9000 workers in an area of about 0.75 km^2. With an extraordinary diversity of built form, street scale and character, day and night time activity, socio-economic mix (although inexorably gentrifying) and capacity to accommodate change over time, the 'first city' example of Kings Cross shows how an entirely 'bottom up' organic process of 'enlightened self-interest' of multiple landowners had to be supplemented by precinct planning and governance at a point where the public interest had to be served. This was in the form of acquisition of private land to create public parks and places and later, a precinct parking station and community facilities. The continuing negotiation and

[2] The Green Square development (http://greensquare.mirvac.com) will achieve a density of approximately 22,000 pp/Ha (61,000 people in 2.78 km^2) once completed. In 2016, Kings Cross-Potts Point had a residential density of 20,600 / km^2 (profile ID and BTS TZN explorer abs.)

Fig. 3.3 Location and aerial view Kings Cross. (source: Simpson)

dialogue between residents, business, the State government and the City of Sydney can be seen as performing a 'curatorial' role in relation to the place and despite the 'messiness' of this networked and heterogeneous governance, it serves to continually recognise and contest the values and qualities of the place. Governance emerged as needed and has enabled adaptation, accommodated increased densities and development, and shifts in activity to occur. Jacobs' four preconditions for good urbanism: density, mixed use, fine grain street pattern and mix of building types and ages (Jacobs, 1961) and other characteristics have been codified by a wide range of theorists and practitioners into a range of metrics and indicators of what makes a place work and 'feel good' (Gehl 2004; Bentley et al. 1985; PPS 2016; CNU n.d), incorporated into statutory planning systems (WAPC and DPI 2009) as sets of metrics in morphological analyses (Salat 2011) and analytic tools such as space syntax (Hillier 2004, 2014), and adopted as a set of key indicators for 'good urbanity'. However, less attention has been given to the political, legislative, governance and land title systems and processes that have led to these places, even though this was central to Jacobs' thinking and critiques.

In 2012 LSE undertook research into the aspects of urban from and governance that provided the capacity for adaptation (Davis and Uffer, 2013). This identified adaptability at multiple scales from the form of the individual building (terrace houses have proved particularly adaptable) to subdivision and street pattern, but also the need for a 'curatorial' role at the precinct scale. Irrespective of whether this role is taken up by government, a private landowner (as is the case with the Grosvenor estate), civil society in the form of neighbourhood and business association it is the principle of subsidiarity and local knowledge, care, responsibility and authority that is key to the 'curatorial' role (Figs. 3.4 and 3.5).

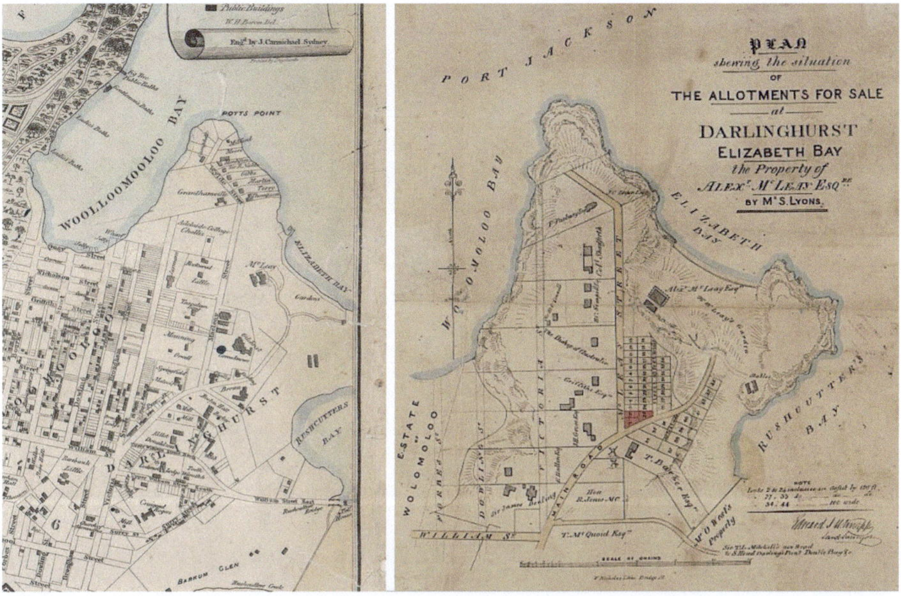

Fig. 3.4 Original land grants and subdivision patterns at different dates. (source: State Library NSW)

Fig. 3.5 Analysis of the resulting grain and diversity of Kings Cross. (source: Johnson, H. Chen, Y. Urbanism Program University of Sydney, Unpublished)

3.4.1 Learning from Kings Cross

Analysing Kings Cross area in Sydney highlights the following:

1. The diversity and 'authenticity' of the area has been enabled by the street and subdivision patterns. These patterns are crucial but often overlooked and should be recognised as a spatial framework and as the basis for individual landowners to respond to shifting opportunities and activities, or for some parts to remain the same- leading to extraordinary juxtapositions. (In general, in Sydney planning controls when subdivision is considered at all, it is usually to encourage site amalgamations to facilitate redevelopment, which is at odds with the fine grain that has been so important);
2. The diversity of built form allowed activities to emerge and be accommodated and provides an engaging urban, active, human scaled aesthetic at street level;
3. The combination of mix of uses, density, permeability of the street and lane patterns, proportion of activated frontage, variety and 'grain' of the subdivision pattern all contribute to 'walkability';
4. The relative high density combined with low car ownership, permeability and accessibility supports local businesses and services;
5. Incremental self-interest produces extraordinary variety, but without an overall master plan or any commitment to public amenity or public good it was left to local government to acquire private land to create public squares and parkland, and provide precinct parking to adapt to changes in transport and activity.
6. Where there is a common interest in character and qualities of the place local groups will emerge spontaneously to take up what may be seen as 'stewardship' or 'curatorial' role. Although these groups may have differing views, even these disputes and discourses may be seen to be a form of resistance to an fettered top-down planning approach.

The key points in relation to resilience that can be derived from the kings Cross example:

- Bottom-up planning processes alone are not sufficient; a curatorial role at precinct scale is required, and
- The process that has formed Kings Cross over a long period of time cannot be replicated, certainly not in a short timeframe, so identifying and requiring crucial spatial arrangements- (subdivision and street patterns) is essential.

3.4.2 Good Urbanism at the Local Scale

Current urban design practice tends to focus on the physical arrangement of the built fabric of the city at a local scale. The 'organic' city characteristics of Kings Cross display Jane Jacobs' four preconditions for good urbanism (Jacobs 1961).

Urban design is awash with codes, rules of thumb, checklists and quantitative analysis in order to define standards, and is prone to applying orthodoxies and attractive 'precedents' generated in very different situations, that act to undermine place and culturally specific responses.

However, without an understanding and conscious engagement with governance and the mechanisms of power and control, that were an essential part of Jacobs' reflections and activism, urban design is rendered superficial, and practice is often limited to a repetition of nostalgic precedents, rote application of orthodoxies and the simulation of authenticity through 'place-making'. The organisation Project for Public Spaces, continuing to draw from the seminal analyses of William H. Whyte (1980), and which might be considered a forerunner in 'placemaking' makes this distinction:

> Placemaking is not the end product, **but a means to an end**. It is the process by which a community defines its own priorities. This is something that government officials and self-proclaimed Placemakers ignore at their own peril. (PPS 2016)

So while these planning documents clamour for 'unique identity', 'contextually responsiveness' and 'authenticity' the irony is that the results are virtually indistinguishable around the world. This should come as no surprise when the normalising and codifying tendencies and needs of neoliberalism and globalisation can be seen to as driving a new 'universalism' far more extensive and pervasive than modernism. It needs to be recognised that the self-organisation and emergence when operating at a global scale, will tend not only concentrate power and wealth, but will also lead to the emergence of particular urban and building typologies and will tend to suppress difference, authenticity and the emergence of local patterns of behaviour and urbanism.

In other words, if it is agreed that the urban characteristics of places like Kings Cross, its ability to adapt, intensify and change over time are very desirable, and that the processes cannot be repeated, then it is important to identify the key spatial elements (diversity in street and subdivision patterns) and key attributes of the governance system that provide the basis for innovation and adaptation (diversity in ownership, activity, power, and their respective layered networks). These in turn can be seen as key to its 'resilience', and important to recognise, actively promote, and protect.

3.5 The 'Functionalist' Second City

More's neologism 'utopia' (More 1516) is an ambiguous but deliberate conflation of 'outopia'– not a place/no place and 'eutopia', a good/better place and it spans the two aspects of the 'utopian project': the conceptualisation of an ideal that does not yet exist, and the attempt to see the existing city from that perspective and remake it in its image. The definition of a singular ideal coherent vision and the attempt to apply it to the existing city is not only at the heart of modernist planning, but also fits comfortably with neoliberal narratives with their focus on productivity, efficiency and the need to provide certainty for investment as outlined earlier (Wright 2013).

The dual aspects of ideal and its application are evident in the County of Cumberland Plan (County of Cumberland 1948), which clearly shows the tension between the ideal and the existing, between the desire and need to 're-direct' the city and the momentum and inertia of the existing physical city and the path dependencies of institutional, bureaucratic, social and political arrangements. The County of Cumberland was a 'special-purpose-planning' organisation created under the Local Government Act. Consequently there are both pragmatic-eutopian aspects, including the dismissal of satellite towns as an unworkable strategy in Australian conditions, and improvements to the road network on which the whole plan was founded (Gibbons 1978; Brewer and Cusack 2006; Department of Main Roads 1945) and idealistic-outopian aspects to the plan; an encircling 'green-belt', a 'green web' of open space, (the former directly transferred from Abercrombie's London Plan and the later emulating the green webs in Stockholm, Helsinki and the Copenhagen 'finger plan') and extensive urban renewal all of which drew directly from the inter-war and post-war programs of urban renewal and reconstruction in Britain (Fig. 3.6).

The plan sets out a range of responses to emerging issues including providing more services and employment closer to residents in the west through a strategy of 'dispersal'.

The 'tension' is clearest in the proposal for broad scale demolition of existing suburbs and reconfiguration of existing centres where the ideal is imposed on the existing fabric in a tidying up exercise applied through zoning, identification of major road reservations through existing urban areas and redevelopment of town centres (Figs. 3.7 and 3.8).

The tension between the ideal and how it might be imposed on an existing city was not a problem of empty expanses of the site of Canberra, the Australian apo-

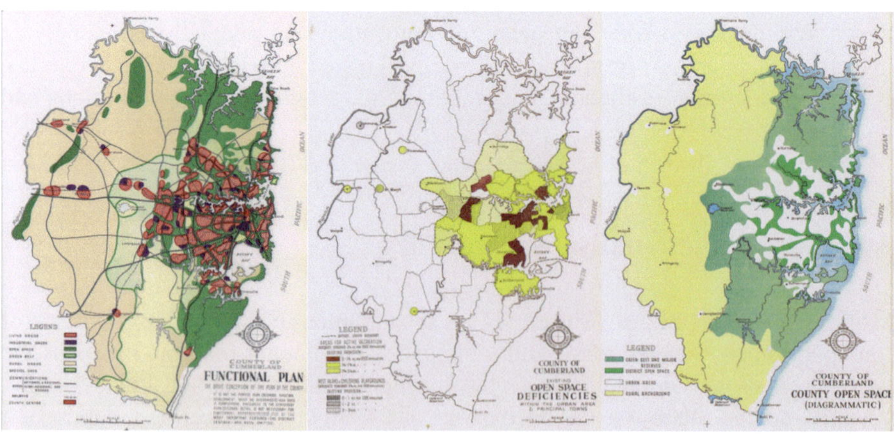

Fig. 3.6 Functional Plan', 'Open Space Deficiencies' and proposed 'County Open Space'. Source: County of Cumberland Plan 1948

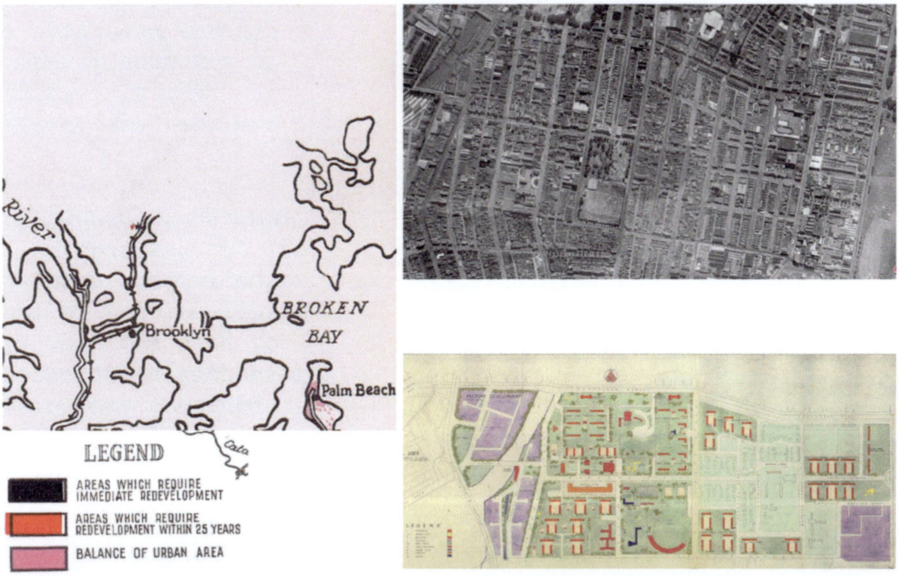

Fig. 3.7 Proposed renewal areas, Conditions in 1943, Proposed redevelopment scheme. (Sources: County of Cumberland Plan, Six maps, City of Sydney Archives)

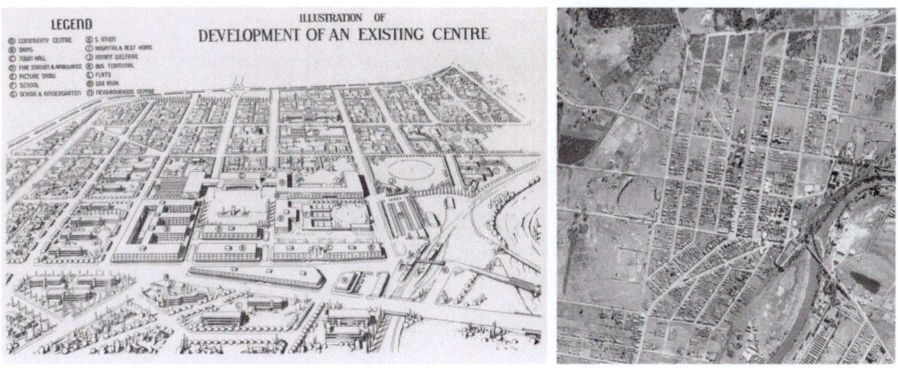

Fig. 3.8 Development of an existing centre (Liverpool) conditions in 1943. (Sources: County of Cumberland Plan, Six maps)

theosis of modernist planning. The attitudes to public domain, technocratic knowledge, power and control, evident in the post war development of Canberra are distinctly modernist. The National Capital Development Commission can be seen as a bureaucracy unfettered by history, existing land-ownings or for that matter, competing government departments. Such a 'tabula rasa' only exists for a short period of time, and vested interests emerge quickly as do attachments to place and private values (Fig. 3.9).

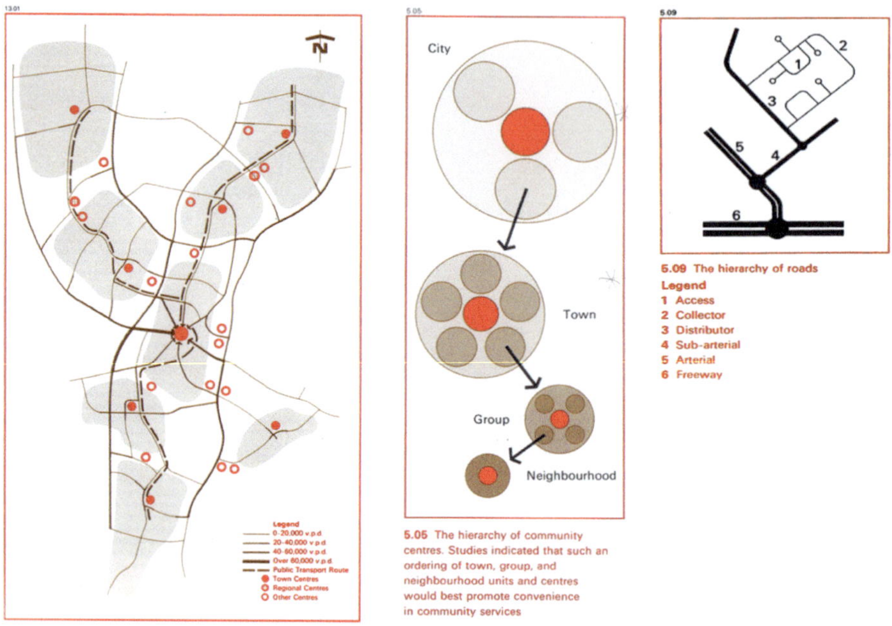

Fig. 3.9 Canberra: a planners paradise: diagrams of hierarchical structures: Urban, Centres, Roads. (Source: Tomorrow's Canberra, NCDC 1970)

In contrast to the path dependencies of bureaucracies and vested interests in the existing cities, landscape values and a deliberate attempt to ensure that speculators would not profit from the investment of public funds were arguably the principal parameters in Canberra (Mead 1911; Brennan 1971; Bourassa et al. 1994). The meaning and interpretation of 'bush capital' and 'garden city', the geometry and different planning approaches have been re-interpreted, appropriated by different groups for different polemics, and consequently adjusted significantly, at a number of junctures. The fact that the bush and landscape play a prominent role in these discourses is significant in itself. With few exceptions (Braddon, Kingston, parts of the Civic), Canberra has failed to achieve the sense of urbanity of the organic first city. The precise relationships between buildings and open space and the public spaces that are critical to a sense of spatial definition and vitality, as imagined by Griffin have not endured as a shared value like the landscape. Urban design in Canberra continues to struggle in this regard where the sheer amount of open space, and overly generous setbacks and separation between buildings is given precedence over spatial definition. In other words: a modernist 'pavilions in space' approach dominates.

However, proposals to densify centres and corridors have the potential to achieve a level of urbanity and accommodate at least 50% of the required additional dwellings required for population growth and demographic change, with the potential to significantly reduce CO_2 emissions to 2030 (SGS 2009). The irony is that the capacity for change is concentrated on the vast setback and underutilised opens spaces in and around the town, 'group' and local centres. The crystalline logic of the 'predict and provide' car-dependent planning that was possible in Canberra did not anticipate the extent of demographic and social changes that have occurred. But the mere fact that there was, and remains a very high proportion of public land, even though the control that could be exercised through this ownership compared to other cities was not exercised, (Neutz 1988) now provides Canberra with its adaptive capacity. Canberra has a great potential to adapt and intensify, but this will only occur in relation to the shared but contested understanding of public values, the bush capital and the landscape settings.

How did this situation; the valuing of landscape and the amount of public land come into being? Arguably it came from the intersection and alignment of the aspirations of the newly constituted commonwealth for a capital that would embody the aspirations and values of the Australian people into the future- a truly 'outopian' ambition, with Griffins' extraordinary ability to embody these values into the landscape and in so doing strengthen both and indissolvably join ethos and landscape.

As put by Ed Wensing (2013):

> Canberra is thus the city that the Federation created on the basis of a host of some of the world's most socially progressive ideas, which were enthusiastically debated in the 1890s colonial parliaments and the first federal parliaments at the time. Canberra, through its leasehold system, its street-naming, its national institutions and its history of founding, egalitarian ideas, is the Australian city which best represents our country's noble history of democratic commitment and achievement

However, as can be inferred from this quotation, the commitment to the landscape and the public good existed at the national level. The Australian Capital Territory does not have full statehood, and ironically as the symbolic and actual locus of Australian democracy, suffer a 'democratic deficit' (Halligan et al. 2002) with less representation than other parts of Australia, while at the same time having a bureaucracy that had the most technocratic origins that were able to exercise the most modernist planning functions and concepts with minimal resistance. However, as residents literally and 'spiritually' took ownership of place, the inadequacies of the top-down technocratic model became apparent. The spontaneous formation of an intermediate level of place-based governance in response to a 'democratic deficit' in the form of 'community councils', which although having not formal statutory role, are supported by the ACT government and act as 'stewards' of their respective suburbs and places. However, even with extensive consultation processes such as 'Time to Talk' (Elton 2012) in the revision of the Territory Plan, with no formal statutory status these organisations can feel sidelined (Stewart and Lithgow 2015) (Fig. 3.10).

Fig. 3.10 Woden: unintended adaptive capacity in open space setbacks and surface car parking. (Sources: Google Earth 2009, SGS, simpson+wilson 2009)

3.5.1 Learning from Canberra

1. A high proportion of land in public ownership and the valuing of landscape have led to a high adaptive capacity. Ironically the spatial extravagance and the closed end-state of the master plans have provided the unintended potential for further development and intensification;
2. The prioritization of public over private values was given effect mainly through the repurchase of private land for the establishment of the capital- effectively a reversal of the original granting of crown land to private interests that in turn was a reflection of the ethos of the time shortly after federation;
3. A sensitive response to landscape is the generator of diversity and identity, both in the relation of developed areas to bushland left largely in its original state and the acceptance responsibility for managing the extensive 'National Capital Open Space System' (NCOSS, NCDC 1983) and the integration of intentional designed and urban landscapes with a mix of exotic and native species:
4. Where there is a 'democratic deficit' or insufficient subsidiarity, informal representative and advocacy groups will emerge to fill the gap;
5. Where the power of the community is circumscribed and limited to particular parts of a technocratic process rather them being empowered to create place, frustration will continue;

6. The explicit initial statement of an 'ethos' of the primacy of the public good and the importance of the landscape, embodied in the original decision to establish a capital for the newly federated nation, set out in the principles and aspirations in the competition brief and made evident in the Griffins' design response, has provided basis for on going argument, justification and rationale used variously by bureaucrats, politicians, planners and the public.

3.5.2 Good Urbanism at the City Scale

Canberra has been very effective in laying out a structure that can accommodate the car but may now also accommodate transit, although the clear spatial logic of the 'Y" plan has been undermined by the release of lands in the Molonglo Valley. The over provision of open space and low density is some areas may have been seen as a waste in the system, but it is this 'redundancy' that now provides Canberra with much of its adaptive capacity for intensification at the city scale.

Achieving 'diversity' at the local scale both in the short and longer term will be more difficult. This is because there is no alternative to the idea of fixed end-state master plans that will not easily accommodate change. Nor are there the governance arrangements (subsidiarity) to allow this to happen. Ironically, as higher densities are planned with the aim of producing a more 'compact' city and to support transit, the built form is maximised within existing codes for solar access, privacy, onsite car parking and so on that leaves much less potential for change (adaptive capacity) than in earlier plans; redundancy has not been recognised as important and has not been allowed for, even though this would be relatively easy to do, given that the ACT government owns the land.

Many newer areas that have been designed in accordance with 'transit-oriented development' (TOD) do not have the sense of place, nor the capacity to change and evolve over time, or the likelihood of meeting Jacobs' four preconditions for urbanity.

Top-down planning alone, that meets all 'standards' for road design, parks and social services will not and cannot produce authentic adaptive places. For that to happen an alternative model that embraces the dialectic between top-down and bottom-up processes, that operates at a very local scale, that responds to the 'democratic deficit' is needed. Rather than suppressing or denying this through conventional consultation on predetermined plans, intermediate levels of governance (subsidiarity) should be encouraged and cultivated as this would have profound implications for planning, urban design, architecture and landscape practice. There is little capacity for innovation if centralised control is retained and government is reluctant to hand over control or to develop alternative forms of governance (Fig. 3.11).

Fig. 3.11 Clash of orthodoxies: modernist road planning meets 'transit oriented development = highest density with lowest amenity. John Gorton Drive, Coombs, Flemington Road, Gungahlin. (Source Googlemaps, Google Street view)

3.6 The Third City

The Western Parkland City; the Third City, has an approximate area of 800 km^2. At its core is ***wianamatta***; South Creek, a winding stream 70 km long, 40 km north to south with a catchment area of 629 km^2. Occupying gentle undulating country more fertile that the sandstone to the east, its forests were comprehensively cleared (Fig. 3.12). Urban expansion westward has bisected the catchment and it has been planned in a piecemeal. It is only recently that the prospect of significant urbanisation of the majority of the catchment that will result in radical changes to the hydrology and potential impacts on the receiving waters of the Hawkesbury Nepean that the catchment has emerged as landscape entity of interest and concern, even though these issues have been appreciated for decades.

The catchment has a wide variety of land uses, and some of the earliest land grants after European invasion were along its fertile banks. Although the earliest grants were on the proviso that the land be put to productive use, by 1848 virtually all the land with the exception of reservations for roads and some 'commons', were in private ownership, with a number of very large consolidated land-ownings, some in excess of 2500 Ha (Fig. 3.13). The speculative acquisition, banking and trading of land as a commodity had overtaken its productive role by the mid 1800s (Else-Mitchell 1974; Spearritt 2004).

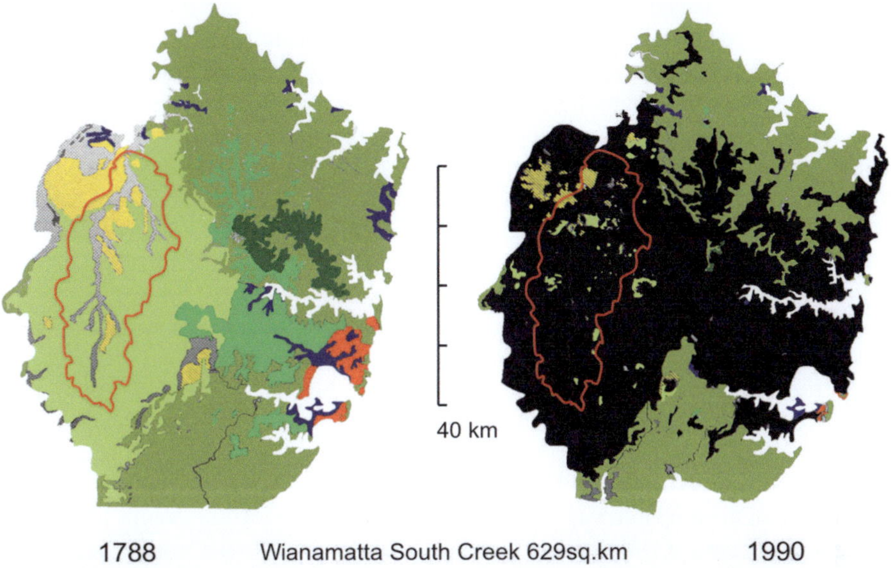

Fig. 3.12 Impact of colonisation, agriculture and urbanisation on vegetation communities on the Cumberland Plain. 1788–1990. (Source: adapted from Benson and Howell 1990)

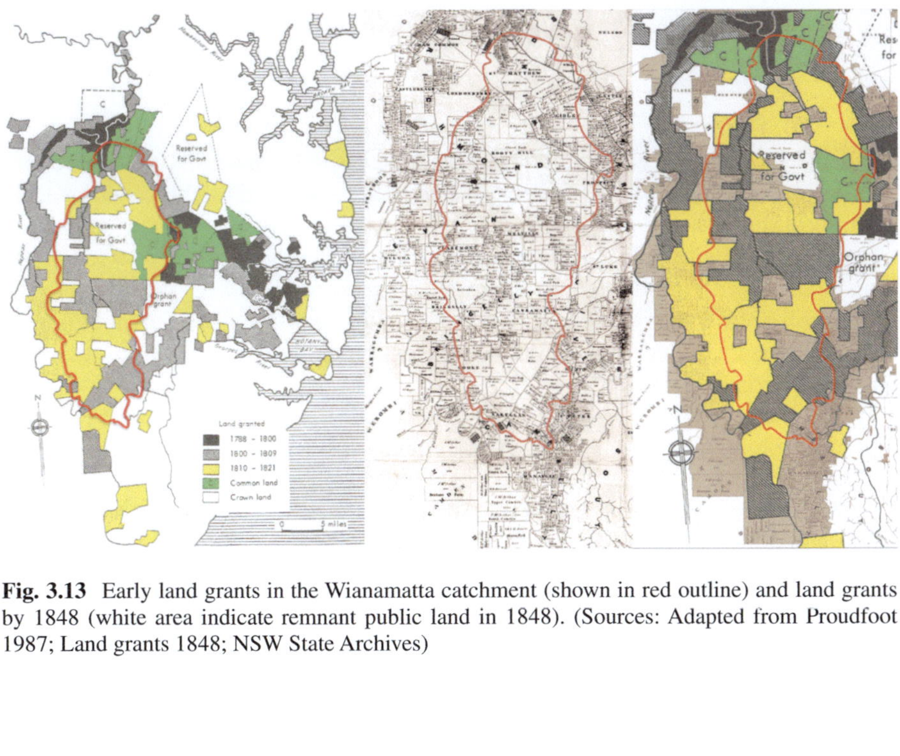

Fig. 3.13 Early land grants in the Wianamatta catchment (shown in red outline) and land grants by 1848 (white area indicate remnant public land in 1848). (Sources: Adapted from Proudfoot 1987; Land grants 1848; NSW State Archives)

The fundamental challenges for the Western Parkland City and for the Wianamatta catchment in particular, are the identification and re-establishment of public values and the public interest over this privatised and highly modified landscape, protection and restoration of natural systems both for their intrinsic and instrumental values. Prior to European land grant expropriation, the entire landscape could be seen as the 'public domain' of the indigenous population. Although eminent domain provisions exist in many countries for public purposes, the difficulty of reclaiming land was seen decades earlier with the collapse of the 'greenbelt' of the County of Cumberland Plan, primarily due to the withdrawal of the Commonwealth from the tripartite and equal funding commitment of the three levels of government to the re-purchasing of land on the one hand and land-owner development expectations on the other (Evans and Freestone 2009; Spearitt 2000; SMH 1951).

As has been shown in Canberra, retention of a high proportion of developable land in public ownership is key to the adaptive capacity of the city and this is evident in many cities around the world and in particularly in urban renewal, such as in Marseille, (Euromediterranee 2018), Stockholm and Helsinki, both of which retain about 70% of developable land in public ownership (Berry 1995; Neutz 1988), and in the renewal area of Hafencity in Hamburg (Clark et al. 2010). Counter-intuitively, the amount of land in public ownership is much higher in the densest parts of the city than the outskirts. New York has 45% of its land in public ownership (Angel 2012, p. 72), in central Paris the proportion of land in public ownership is higher than outside the *peripherique* (APUR 2013).

3.6.1 Principles for the Third City

From the preceding discussion, two principal aspects for planning the future city can be identified: ethos and landscape. These are linked; the ethos encompasses both an attitude to the public good and to how that plays out in the landscape. Without lapsing into physical determinism, if it is accepted that there is ***any*** relationship between the way people interact with each other and the place where the interaction occurs, then the extent to which the spatial organisation, the urban form and structure facilitates some patterns and inhibits or impedes others is the core concern of urban design.

If the relationship between the ethos and the landscape is seen as dynamic, shifting and evolving as the multitude of individuals' decisions and actions to inhabit and modify that landscape, then thinking about the elements that remain relatively fixed is important (streets and to a lesser extent subdivision, and lesser still, buildings).

Furthermore, this interaction is a sufficient basis for a place specific 'ethos' to emerge and for identity and authenticity, so sort after, but so lacking in much greenfield expansion to develop. This place specificity may be seen as a viable and enduring form of resistance to the forces of normalisation that accompany globalisation.

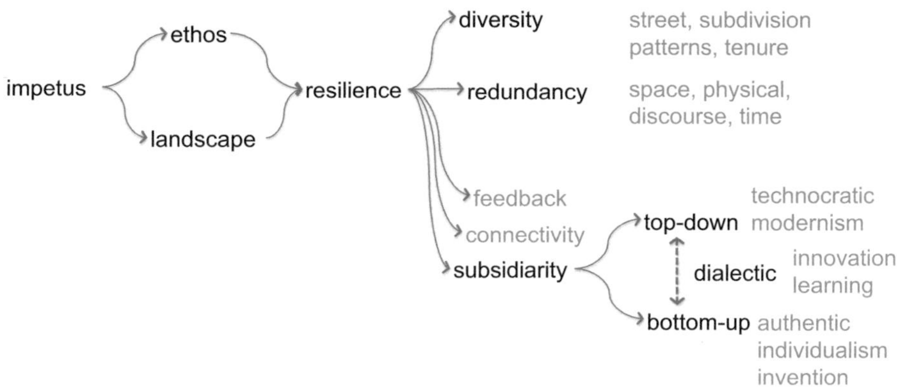

Fig. 3.14 Resilience principles applicable to urban design: process and product. (Source: Simpson)

It follows that the initial 'policy settings' and articulation of values are crucial. Even though these will evolve and be contested over time, as has happened in Canberra, these clear statements of the intrinsic and instrumental values of the landscape and the prioritisation of the public domain, public interest and public land may be seen as the key parameters for how the land is occupied and transformed. However these 'policy settings' but are generally overlooked by the juggernaut of urban expansion.

The relationship between ethos and landscape, resilience and the various aspects of resilience that relate to urbanity and urban design, such as diversity, redundancy, and subsidiarity, are show on the diagram below. This provides a framework for thinking about how the planning for the western parkland city could proceed based on the simple proposition: the good city is a resilient city.

The diagram (Fig. 3.14) shows the dual nature of subsidiarity governance as a dialectic, rather than a hierarchical relationship. This relationship provides the basis for experimentation and learning- with top-down and bottom-up being interdependent but with distinct roles.

The starting point; 'impetus' is the desire for the city to continue to exist, and the reason all cities exist is a combination of a rationale (economic or social) – 'ethos' – and a location – the landscape.

3.6.2 Ethos

Ethos is a dynamic social construct. It is usually implicit, and can generally only be deduced and observed second hand through its effects, rather than though its explicit expression.

It is often only at historic junctures that there is a deliberative process to define and express the *'zeitgeist'* and ethos of a place and people. The obvious catalysts would be the framing of a national constitution, or as has been alluded to, the framing of the competition for the design of the National Capital, the land tenure system and attitude to land in Canberra. But it may also be stimulated by events such as the 2000 Olympics in Sydney. The conceptualisation of the Third City has a similar potential, and the starting point is the attitude to land and landscape.

In the past the definition of 'ethos' has emerged or been based on ideological, political and cultural values, which are contested in the political sphere, hence based solely on power relations. As an alternative 'resilience' can act as both a unifying and evidence-based framework for defining an ethos that is less subjective and more conciliatory: it reconciles the need for economic growth with social justice and with protection of the environment by recognising their interdependencies. In other words, it is consistent with the concept and thinking about socio-ecological systems and recognising them as complex adaptive systems (Biggs et al. 2015).

3.6.3 Landscape

Most of the existing city's eastern urban area discharges wastewater to the ocean and is blessed with higher rainfall, cooling breezes and extensive remnant vegetation in unbuildable areas. The Western Parkland City is in a different climate zone, has very different weather, is subject to many more days of extreme heat over 35 C, and will have to manage wastewater and significantly increased storm-water flows within the Wianamatta catchment. The landscape will have to be reconstructed, re-vegetated and irrigated to mitigate higher temperatures and make the place both attractive and liveable whilst protecting the natural quality of the creek itself and the ecosystems in the Hawkesbury River into which it flows.

The term 'parkland' is intended to convey the idea that this devastated landscape will have to be re-made; it is not simply a matter of identifying and protecting remnant habitat, it must be subject to human intention in a more deliberative and integrated way than has been the case. Although it may be underpinned by environmental protection policies for aquatic and terrestrial ecosystems, and endangered communities and species, it goes far beyond these fragmented pieces of legislation by necessity.

In the *anthropocene* more generally, and in Wianamatta in particular, nature cannot look after itself.

Although the shift westward has be driven by the desire to rebalance the economy and to provide more social, economic and cultural opportunities for existing and future population of western Sydney, moving into a very different landscape is beginning to drive a recognition of the city as a interdependent 'socio-ecological' system out of a necessity that has not been so immediate or pressing. In the west, the massive biological and thermal sink of the South Pacific Ocean is not as accessible, and the behaviour of the airshed means air quality and extreme heat are pressing issues.

Every Australian city needs a 'landscape framework' – a combination of scenic hills, watercourses, remnant vegetation and biodiversity, and strategies to mediate climate extremes with vegetation. Unlike Canberra, with the National Capital Open Space System (Seddon 1983) NSW relies on siloed legislation relating to flooding and storm-water management, biodiversity, riparian vegetation, heritage and recreation without an overall appreciation or framework for planning and managing the landscape.

In response, open space is being reconceived as 'green infrastructure' and is on the cusp of being considered 'essential infrastructure', justified and argued for on its instrumental value. The 'green grid' (GANSW & Tyrrell Studios 2016) proposes an interconnected regional open space system that integrates recreational and habitat values.

At the same time, and for the first time in any metropolitan plan for Sydney, it is prefaced not only by an 'acknowledgement of country' but by recognition of aboriginal values, stewardship and the desire to have those values drive planning: 'if we care for country, country will care for us' (GSC 2018,1). Planning for the Third City has the opportunity of developing a comprehensive landscape framework that embodies indigenous, habitat, infrastructure, aesthetic and recreational values.

The extensive Wianamatta, with its adaptive, mediating and compensatory flood plains, could be the trunk of a network of open space connections to provide a comprehensive alternative way of getting around the third city and to all surrounding centres. In this landscape the highest density should take advantage of the best locations in the landscape, such as the top of hills or close to the green amenity of the watercourses in order to benefit as many people as possible.

In other words the ethos of the Western Parkland City may be seen as both pragmatic/instrumental and aspirational/cultural at the same time- the ethos must be an explicit expressions of both what is needed and what is desired for this socio-ecological system.

3.6.4 Diversity

Human systems operating at the global scale tend towards large monocultures (Ricoeur 1961; Schafer 2001) and concentration of power and wealth (Vitali et al. 2011). In other words they tend towards a reduction in diversity. This is a fundamental characteristic of capitalism, and distinct from the diversity in natural systems that arises primarily from the subtle variations in place and context (ecological niches) driving increased speciation. In contrast the combination of globalised finance, standards, typologies (e.g. 'hypermarts') generic consumer preferences and the technological capacity to deny place: either by obliterating it (e.g. terraforming for motorways, ports and airports) or being able to ignore it (e.g. air conditioning) can be seen as a form of self-organisation and emergence that works against diversity.

Existing diversity needs to be protected as a source of invention, a place for trial and error in urban experiments, which may or may not become 'emergent' – i.e.

'innovation being incorporated in the mainstream'. Once mainstreamed other sources of diversity must be found for invention and 'disruptive innovation'. It follows that if diversity is valued, both for its intrinsic 'interest' and 'variety' in the urban context, and its instrumental value in resilience then in human systems there must be active policies to protect and propagate diversity- because our globalised capitalist system will work to eliminate it.

An attitude to landscape must be conscious and clearly stated as a prime objective in any urban design. Sensitivity to the variation in landscape and climate can generate diversity, identity and authenticity, and urban quality benefits of adherence to these principles of resilience. The more subtle the variations in the landscape the more sensitive and nuanced urban responses must be.

3.6.5 Redundancy

In relation to the physical layout of cities and urban design plenty of public land is a key component in this framework. Through a 'resilience lens', layered governance ample public space and facilities and a substantial proportion of land retained in public ownership are not seen as wasteful or inefficient, but this 'unplanned space (Roggema 2012) rather provides the capacity for the city to evolve and adapt. Having sufficient 'space': physical, in timeand in discourse is a way of 'accommodating' uncertainty and allowing invention and experimentation.

3.6.6 Subsidiarity and Authenticity

The application of resilience principles is very different to the hierarchical application of a 'Master Plan' where there is a cascading and nested level of detail that provides a technocratic 'line of sight' from the global to the local. In urban change this is not and has not ever been possible because it assumes a single definable end state. Instead, all that can be hoped for is an evaluation of whether something is more or less good in relation to a certain goal, and even in assessing these it is essential that this is checked at higher and lower scales.

Resilience is an effective form of resistance to global tendencies. Global capital does not care whether a city fails or not; there will be others in the same league to invest in. Capital has already hedged its bets and on-sold its risk: the only people who are truly interested in the success/persistence/sustainability of one city over another are the citizens themselves.

All cities undertake metropolitan planning, and in this very act attempt to define the city as an object that is characterised, personified, understood and programmed. The dominant narrative in these plans speaks about their respective cities as knowable 'entities'. Much of the literature stemming from Sassen's conceptualization of 'global cities' (Sassen 2001), the emergence of various 'liveability' indexes, checklists of ingredients for 'creativity' (Florida 2002; Montgomery 2005), the celebration of cities a 'triumphing' over other settlement patterns (Glaeser 2012), as the most appropriate locus of addressing climate change (Hurst and Clement-Jones 2016; ICLEI undated; COP23 2017) and human well being (OECD 2017) and resilience in the face of natural disasters and mega trends (UNISDR undated; 100Resilient Cities undated). All of these formulations push towards a generic understanding of cities as 'urbanisation' that tends to look for similarities rather than difference. While it may be true that the affluent in every city have more in common with each other than with the disadvantaged, the solution to this does not lie in top-down reform but allowing bottom-up emergence.

The differences that do exist give places traces of authenticity, but are subsumed within this generic recipe for the successful global city. Chinatowns, historic quarters and more generally conserving 'heritage districts' and 'fine grain centres', are reduced to nothing more than essential elements. As if these were theme parks, that every city should have at least one example of to be competitive and provide 'destinations' both for citizens within the city and from outside. This dismal view of the city, where even every 'historical centre' becomes a blur of tourist shops, restaurants, and museums was complemented in Koolhaas' idea of the 'generic city' (Koolhaas 1995) comprised of urban forms and built typologies of malls, suburbia, industrial parks and logistic estates that are known, understood, commodified, securitised and propagated by international finance, looking for a secure place to land. This process of generating league tables and 'formulas for success', in itself has come to be the predominant way cities are understood, planned and conceived. The aversion or disinterest of economists to consider the distributional and social effects of economic activity, and the almost nihilistic and anti humanist capitulation of architecture to the needs of international capital (*starchitecture*, iconic bling towers piercing the skyline are simply the complement to fake authenticity with often distressed paint, sandblasted timber, rusted metal at ground level) can be seen as a consistent and universal end game played out globally.

3.7 The Good City Is the Resilient City

The desire to have face-to-face contact, whatever the technological, or economic context is likely to remain, and will not change within the foreseeable planning horizons. AI, autonomous vehicles and IoT will emerge but these need to be directed to achieve human wellbeing, that they will be can't be taken for granted (Hariri 2017). They may augment reality but they will not and should not replace it. Urbanity is the condition that has arisen from this desire for face-to-face contact, and through the accompanying redirection and concentration of resources has allowed the human population to grow to where it is now. It is easy to argue that is it good that urbanisation only occupies 2.5% of the planets surface.

With urbanity comes a necessary 'civility' and learning the skill to deal with strangers, which is the basis for cosmopolitanism (Bauman 2006, 2007). So the condition in the third city is likely to be the unlikely combination of cosmopolitanism that has been associated and almost synonymous with displacement and placelessness in the past, with a profound connection to the landscape and place in the third city. In the 2016 census 40% of Sydney's residents were born overseas, and this is likely to increase. (ABS 2016).

The vision for the third city is not a collection of enclaves, but is open ended and continually reimagined and renegotiated. Socio-ecological systems that constitute the city are interdependent. The separation of social and ecological systems reduces resilience. The complexity of the city, exacerbated by the uncertainties of the future, has been characterized as a wicked problem (Rittel and Webber 1973; Conklin 2001). There is no 'solution' to such a problem, there can only be trial and error attempts to make the system work better. But neoliberalism and managerialism considers local governance to be an impediment, which may be true in its own terms but not in achieving broader social and environmental outcomes (Buxton et al. 2012; Gleeson et al. 2012). An alternative approach would recognise that major transport infrastructure does indeed change space and time. A distinction between strategic structuring infrastructure and 'following infrastructure is useful (Spiller et al. 2015; Simpson 2015), however this is still within the modernist technocratic view of the city, where predict and provide dominate and the importance of place and the experience of the individual citizen is still overlooked.

3.8 The Role of Urban Design

There have been many attempts to transfer well-recognised aspects of resilience to design. Such approaches may produce a simulacrum of resilience, but by not considering multiple scales, both physical and organizational, this is doomed to fail. The city is a *politico-economic-socio-ecological* system, and it is impossible to reduce this system to the spatial aspects or checklists of attributes that are usually the focus and extent of urban design. This has happened in architecture in relation to sustainability, where checklists of sustainable building, such as LEED (USGBC undated) and BREEAM (BREEAM undated) measure the supposed sustainability of buildings, when they exist in a fundamentally unsustainable context. However, urban design is better equipped than architecture to extend its influence into the social, political and economic domains, and be recast as 'urbanism'.

A possible way to deal with this apparently overwhelming complexity is to see the city as a mosaic of local places, with their own constituencies, potentials and challenges, sitting within an overall spatial, infrastructural, learning governance framework. The recent Greater Sydney Regional Plan puts great emphasis on place based planning, precincts, 'collaboration areas' as a means of dealing with the and planning for these 'tiles' in the urban 'mosaic' is essentially a bottom- up process, that would ideally engage with local interests, values and appreciation of place. The role of the urban designer becomes an intermediary in a 'dialectic' process;

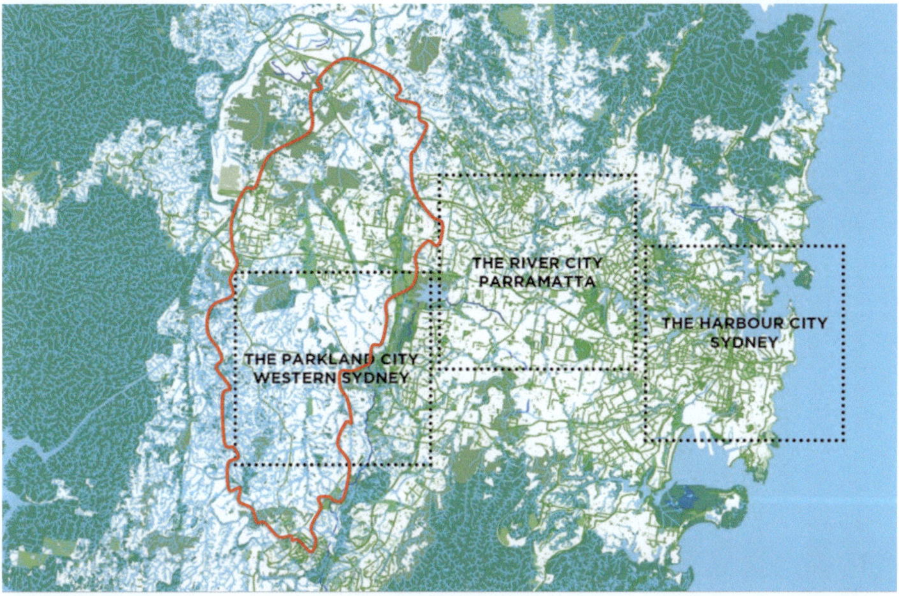

Fig. 3.15 A metropolis of three cities areas: landscape and Wianamatta. (source: Greater Sydney Commission 2018a, b; GANSW & Tyrrell Studios 2016)

understanding the 'top –down' spatial and policy framework, and how this might be inhabited and exploited by local initiatives.

This is very different to the vision set out in the urban design protocol for Australia (ASBEC 2011), which still infers a hierarchical 'line of sight' 'flow of strategic intent' (Fig. 3.15). This model still pervades metropolitan planning in Sydney with the idea of 'housing targets' that are given to local government to ensure 'performance' (Troy 2016).

It is clear that current planning processes are nowhere near the top of Arnsteins ladder of participation (Arnstein 1969), but Arnstein does not distinguish the appropriateness of different modes of participation in relation to scale. There is still a need for overarching spatial frame and city shaping infrastructure, and consultation, engagement processes for these will be very different, and serve a different purpose to engagement and participatory 'co-design' processes that may be appropriate and possible to local planning.

In part, the desire for a single overarching 'narrative' for the city can be seen as the natural accompaniment to the certainty that businesses demand as part of neoliberal privatisation of public services and infrastructure.

When the primary aim of ensuring the wellbeing of citizens is substituted by international competitiveness, and global city rank, it follows that 'bottom-up' processes are seen as unnecessarily time consuming and an 'impediment to city-making'.

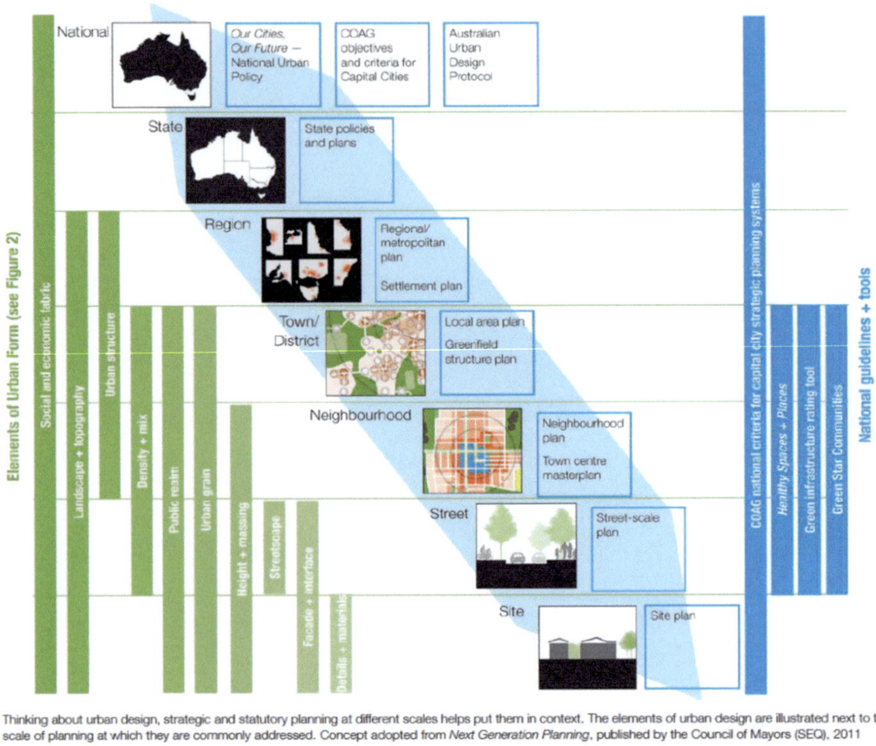

Fig. 3.16 Idealised 'line of sight' in planning and urban design that suggests perfect information and transmission of intent without an indication of learning or innovation dynamics or adaptive urban management. (Source: Urbandesign.org.au 2011)

As has been seen in many instances, in Sydney and Canberra, this is a false economy and actually less 'efficient' more times consuming, divisive and costly than an embracing of 'bottom-up processes'. Additionally, the co-benefits' of multi-layered governance, 'stickyness' of social interactions that are fundamental to resilience are not realised (Fig. 3.16).

3.9 Conclusion

Learning from the first and second city is not a matter of physical determinism or nostalgia. The first and second city examples have demonstrable performative benefits, in other words there are 'universal principles' related to both physical and spatial arrangements, and to governance that can be observed to have contributed to the adaptive capacity of these places over time.

The design of future cities, such as the Third City in Sydney, can learn from the technocratic urbanism of Canberra only because it was accompanied by a clear statement of an 'ethos' that put the public interest and landscape values first. It can also draw from the organic urbanism of Kings Cross while acknowledging that incremental self-interest is not enough; while the subdivision and street pattern has generated an extraordinarily rich character, its ability to adapt has required the provision of public precinct scale infrastructure and retrospectively establishes a public domain. In both Kings Cross and Canberra intermediate forms of governance have emerged spontaneously as inhabitants and businesses have taken 'ownership' of the place, demanding 'agency' that has resulted in a layered stewardship that displays the characteristics of subsidiarity. In other words; 'democracy will out' where there is a democratic deficit, so why not embrace and cultivate it from the outset?

Planning for a Resilient Third City should emphasise diversity, redundancy and recognise that subsidiarity in governance is not only effective but will arise in any case: that both top-down and bottom-up processes are needed, and these are in a dialectical relationship not a hierarchical one. Having sufficient 'space': physical, in time and in discourse is a way of 'accommodating' uncertainty and allowing invention and experimentation; as noted, having spatial and organisational 'redundancy' is not 'wasteful' or inefficient, when seen through a lens of resilience.

Thinking about the principles for the 'third' city is both prospective and reflective, and provides the basis for remaking the existing city as much as designing the new and future city. Contrary to the universalist hegemonic standardising tendencies of modernism, now 'turbo-charged' by globalised neoliberalism, the principles formulated for the third city, founded on 'resilience' can also be used to revisit and remake existing cities, in Sydney or elsewhere and may be seen as a form of resistance to these forces.

References

100Resilient Cities (undated) What is urban resilience? URL: http://www.100resilientcities.org/resources/. Accessed 27 January 2018

ABS (2016) http://www.censusdata.abs.gov.au/census_services/getproduct/census/2016/quickstat/1GSYD. Accessed 11 March 2018

Angel S (2012) Planet of cities. Lincoln Institute of Land Policy, Cambridge, MA

APUR Atelier Parisien d'Urbanisme (2013) Atlas du grand Paris, p 34

Arnstein SR (1969) A ladder of citizen participation. J Am Plan Assoc 35(4):216–224. https://doi.org/10.1080/01944366908977225

ARUP (2014) City resilience index: research report volume I: desk study

ARUP (2015) Cities resilience index (brochure, undated). https://www.arup.com/projects/c/city-resilience-index

ASBEC (2011) Creating places for people. An urban design protocol for Australian cities. URL: https://urbandesign.org.au/content/uploads/2015/08/INFRA1219_MCU_R_SQUARE_URBAN_PROTOCOLS_1111_WEB_FA2.pdf. Accessed 8 January 2018

Australian Government (2017) http://westernsydneyairport.gov.au. Accessed 27 January 2018

Bauman Z (2006) Liquid fear. Polity Press, Cambridge

Bauman Z (2007) Liquid times: living in an age of uncertainty. Polity Press, Cambridge

Benson D, Howell J (1990) Taken for granted: the bushland of Sydney and its suburbs. Kangaroo Press/Royal Botanic Gardens, Sydney

Bentley I et al (1985) Responsive environments. Architectural Press, London

Berry E (ed) (1995) European cities, planning systems and property markets. Spon, London
Biggs R, Schlüter M, Schoon ML (eds) (2015) Principles for building resilience: Sustaining ecosystem services in social-ecological systems. Cambridge University Press, Cambridge
Bourassa S, Neutze M, Strong A (1994) Leasehold policy and land use planning in Canberra: a critical assessment – Urban Research Program Working paper no. 44, Research School of Social Sciences Australian National University Canberra
BREEAM (undated) What is BREEAM? URL: https://www.breeam.com. Accessed 27 January 2018
Brennan F (1971) Canberra in crisis: a history of land tenure and leasehold administration. Dalton, Canberra, pp 22–35
Brewer J Cusack S (2006) The Sydney orbital – around in circles or moving Sydney forward? 22nd ARRB conference – research into practice, Canberra, Australia
Buxton M, Goodman R, March A (2012) Planning systems, urban form and housing. In: Tomlinson R (ed) Australia's unintended cities. The impact of housing on urban development. CSIRO Publishing, Clayton
Clark G, Huxley J, Mountford D (2010) Organising local economic development, Local Economic and Employment Development (LEED) program. OECD, Paris
CNU Congress for New Urbanism., various documents on line. https://www.cnu.org/resources/tools, http://buildabetterburb.org/tools/
Conklin J (2001) Wicked problems and social complexity, p 11. CogNexus Institute. URL: http://cognexus.org/wpf/wickedproblems.pdf. Accessed 13 December 2017
COP23 (2017) COP23 comes home to Bonn. URL: http://www.cop-23.org. Accessed 28 January 2018
County of Cumberland (1948) County of Cumberland Planning Scheme. County of Cumberland
Davis J, Uffer S (2013) Evolving cities. Exploring the relations between urban form resilience and the governance of urban form. London School of Economics and Political Science. London. URL: https://lsecities.net/objects/research-projects/resilient-urban-form-and-governance. Accessed 8 January 2018
Department of Defence(2016) 2016 Defence White Paper. URL: http://www.defence.gov.au/whitepaper/, https://www.fpwhitepaper.gov.au. Accessed 8 January 2018
Department of Main Roads NSW (1945) Main Road Development Plan
Else-Mitchell R (1974) Legacies of the nineteenth century land reformers from Melville to George John Murtagh Macrossan Lecture
Elton Consulting. (2012) Consultation outcomes report ACT planning strategy. http://www.planning.act.gov.au/__data/assets/pdf_file/0010/894979/Elton_Consultation_Outcomes_Report_May_2012.pdf
Euromediterranee (2018) http://www.euromediterranee.fr/topics/accommodations/la-mixite-au-coeur-de-la-politique-de-logement.html. Accesses 27 January 2018
Evans C, Freestone R (2009) Green regional design: philosophies, policies and products in the evolution of metropolitan open space in Sydney 1948-2008, SOAC 2009. http://soac.fbe.unsw.edu.au/2009/PDF/Evans%20Catherine_Freestone%20Robert.pdf. Accessed 27 January 2018
Feliciotti A, Romice O, Porta S (2016) Design for change: five proxies for resilience in the urban form. Open House Int 41(4):23–30
Florida R (2002) The rise of the creative class. Harper Collins, New York
Friedmann J (2010) Place and place-making in cities: a global perspective. Plan Theory Practice 11(2):149–165
GANSW & Tyrrellstudios (2016) Sydney green grid spatial framework and project opportunities. Department of Planning and Environment, Sydney
Gehl J (2004) Public spaces, public life. Danish Architectural Press, Copenhagen
Gibbons R (1978) Transport planning and administration in Sydney MEcon thesis. University of Sydney

Glaeser E (2012) The triumph of the city. Penguin Random House, London/New York
Gleeson B, Dodson J, Spiller M (2012) Governance, metropolitan planning and city-building: the case for reform. In: Tomlinson R (ed) Australia's unintended cities. CSIRO Publishing, Collingwood
Greater Sydney Commission (2017) Our greater Sydney 2056 – a metropolis of three cities – connecting people. State of New South Wales, Sydney
Greater Sydney Commission (2018a, 1) A metropolis of three cities; Greater Sydney regional plan. State of New South Wales, Sydney
Greater Sydney Commission (2018b, 1) Western district plan. State of New South Wales, Sydney
Halligan J, Skelly L, Wettenhall R (2002) Patterns of representation and community governance in the city state. Canberra Bull Pub Adm 103:47–59
Harriri I 2017 Homo Deus
Hillier W (2004) Space is the machine. UCL, London
Hillier W (2014) Space syntax as a theory as well as a method. http://isuf2014.fe.up.pt/Hillier.pdf
Holling CS (1973) Resilience and stability of ecological systems. Annu Rev Ecol Syst 4:1–23
Hurst T, Clement-Jones A (2016) Deadline 2020. C40 and ARUP. URL: http://www.c40.org/researches/deadline-2020. Accessed 28 January 2018
ICLEI (undated) Low-carbon, climate resilient cities. URL: http://www.archive.iclei.org/index.php?id=800. Accessed 26 January 2018
Jacobs J (1961) The death and life of great American cities. Random House, New York
Koolhaas R (1995) The generic city. In: Koolhaas R, Mau B, Sigler J, Werlemann H (eds) S, M, L, XL. Monacelli Press/010 Publishers, New York/Rotterdam
Mead E (1911) What Australia can teach America. The Independent 71(3272):367–370
Montgomery J (2005) Beware 'the creative class'. Creativity and wealth creation revisited. Local Econ 20(4):337–343
More T (1516) Utopia. Penguin Classics, London
NCDC National Capital Development Commission (Seddon, G) (1983) National Capital Open Space System, Policy Plan and Development Plan
Neutz M (1988) A tale of two cities. Public Land Ownership in Canberra and Stockholm. URU working paper no. 3
Newman P, Beatley T, Boyer H (2009) Resilient cities; overcoming fossil-fuel dependence. Island Press, New York
NSW State Archives and Records (undated) Land grants guide 1788-1856. URL: https://www.records.nsw.gov.au/archives/collections-and-research/guides-and-indexes/land-grants-guide-1788-1856. Accessed 8 January 2018
OECD (2017) How's life? 2017. Measuring well-being. OECD Publishing, Paris. https://doi.org/10.1787/how_life-2017-en
PPS (2016) People for public spaces. https://www.pps.org/article/grplacefeat. Accessed 19 December 2017
Proudfoot H (1987) Exploring Sydney's West. Kangaroo Press, Kenthurst
ResilientCity (2012) https://www.resilientcity.org/index.cfm?pagepath=Resilience/Resilient_Design_Principles&id=11900. Accessed 27 January 2018
Ricoeur P (1961) Universal civilization and national cultures. In: Ricoeur (ed) History and truth. Northwestern University Press, Evanston, pp 276–283
Rittel H, Webber M (1973) Dilemmas in a general theory of planning. Policy Sci 4:155–169. Elsevier Scientific Publishing Company, Inc., Amsterdam. (Reprinted in Cross N (ed) Developments in design methodology. Wiley, Chichester, 1984, pp 135–144)
Roggema R (2012) Swarm planning; The development of a methodology to deal with climate adaptation. PhD-thesis. Delft/Wageningen: Delft University of Technology and Wageningen University and Research Centre. https://doi.org/10.4233/uuid:629daeaa-9975-4ec4-a5a0-23e6714cd35e
Salat S (2011) Cities and forms. CSTB, Hermann Editeurs, Paris

Sassen S (2001) The global city. Princeton University Press, New York/London/Tokyo/Princeton
Schafer W (2001) Global civilisation and local cultures. Int Sociol 16(3)
SGS Economics and Planning, Simpson+Wilson, Kinesis (2009) Urban principles review. ACTPLA, Canberra
Simpson R (2015) Transformational infrastructure: remaking the city and questioning current practice. New Planner 103:18–21
SMH (1951) Time for decision on master plan. Sydney Morning Herald. 6 January 1951. https://trove.nla.gov.au/newspaper/article/18194117. Accessed 27 January 2018
Spearitt P (2000) Sydney's century. UNSW Press, Sydney
Spearitt P (2004) Rum Corps to white-shoe brigade. Griffith Review Edition 2: Dreams of land, Griffith University Press, Brisbane
Spiller M, Vreeburg JHG, Leusbrock I, Zeeman G (2015) Flexible design in water and wastewater engineering – definitions, literature and decision guide. J Environ Manag 149(0):271–281 https://doi.org/10.1016/j.jenvman.2014.09.031
Stewart J, Lithgow S (2015) Problems and prospects in community engagement in urban planning and decision-making- three case studies from the Australian Capital Territory. Policy Stud 36(1):18–34
Stiglitz JE (2015) The great divide: unequal societies and what we can do about them. W.W. Norton, New York
Troy L (2016) Projections, targets, forecasts and models how do Sydney's metro strategies try to anticipate our future housing needs? City Futures Blog. URL: http://blogs.unsw.edu.au/cityfutures/blog/2016/03/projections-targets-forecasts-and-models-how-do-sydneys-metro-strategies-try-to-anticipate-our-future-housing-needs/. Accessed 23 January 2018.
UNISDR (undated) Making cities resilient. URL: https://www.unisdr.org/we/campaign/cities. Accessed 28 January 2018
USGBC (undated) Better building is our legacy. LEED is green building. URL: https://new.usgbc.org/leed. Accessed 27 January 2018
Vitali S, Glattfelder JB, Battiston S (2011) The network of global corporate control. PLoS ONE 6(10):e25995. https://doi.org/10.1371/journal.pone.0025995
Walker B, Salt D (2006) Resilience thinking. Island Press, Washington
Wensing E (2013) Nominating Canberra for National Heritage Listing (text, powerpoint and AHC map). Presentation to the ACT and Region Annual Australian Heritage Partnership Symposium Saturday 20th July 2013. URL: https://openresearch-repository.anu.edu.au/bitstream/1885/10300/1/WensingNominatingCanberra2013Paper.pdf. Accessed 8 January 2018
Whyte WH (1980) The social life of small urban spaces. Project for Public Spaces, New York
Wilkinson R, Pickett K (2011) The spirit level: why greater equality makes societies stronger. Bloomsbury, New York
Wright I (2013) Are we all neoliberals now? Urban Planning in a neoliberal era. '49th ISOCARP Congress. http://www.isocarp.net/Data/case_studies/2412.pdf. Accessed 27 January 2018

Chapter 4
Urban Design in the Fragmented City

Peter Bishop

Abstract Traditional approaches to city planning are being challenged by changes in the role of city government and expectations of citizen involvement. Incremental and tactical strategies are being pioneered as solutions to managing urban change. This poses significant questions on the relationship between city government, professional practice and universities and the ways in which we teach the next generation of urban practitioners.

4.1 Politics and Dark Matter

Physicists have hypothetised that invisible forms of matter and electromagnetic force, dark energy and dark matter, constitute 95.1% of the known Universe. They cannot be observed, but their existence can be deduced through their effect on the 4.9% of the universe that we can see and measure. There is a striking analogy here with the cities and places that we inhabit. They are physical and material; we can see them, use them, walk through and live in them. We can debate their qualities and theorise about them. We can deduce that they came into existence through, and are held in place by, powerful forces. These forces though are largely invisible and difficult to measure. The dark energy that lies beneath the surface of our cities; that creates, adapts, destroys and renews them is in the broadest sense politics. The politics of individuals, communities and organisations work in tandem or in opposition to shape the places in which we live and work.

This chapter explores the way in which recent changes in society and economies are influencing political systems and the ways in which we can best shape and design our cities. If urban planners, designers and architects fail to engage appropriately in the political processes that now operate in cities, their work is likely to become marginalised, unimplemented and irrelevant – or worse cause damage to those communities that must live with the consequences of this failure.

P. Bishop (✉)
Bartlett School of Architecture, University College London, London, UK

R. Roggema (ed.), *Contemporary Urban Design Thinking*, Contemporary Urban Design Thinking, https://doi.org/10.1007/978-3-319-91950-8_4

The chapter starts with an examination of our new world of fragmented certainties and considers the implications for city government. It then examines some of the ways in which practitioners can develop effective strategies to shape cities, and explores the implications for the role of universities both as educators and as research partners for city government. It draws on a range of examples, from London and other cities.

4.2 Planning in an Uncertain World

The Polish sociologist, Zygmunt Bauman argues that in the last 40 or 50 years we have moved from a phase of what he terms 'solid' modernity to a 'liquid' phase (Bauman 2004). 'Solid' modernity was based on a belief that it was possible to make a 'fully rational perfect world'. Change was seen as temporary, and it was only a matter of acquiring enough information, knowledge and technical skills to construct a world that was 'completed' and did not require further development. Solid modernity involved removing unknowns and uncertainties through control over nature and through the creation of hierarchical bureaucracies, rules and regulations. These control mechanisms would remove personal insecurities and would make human life well ordered and familiar. Planning strategies based on precedent and the steady accumulation of knowledge made sense in such a world of relatively slow change.

Bauman contrasts this period with the present day. Over the past 30 years we have seen the collapse of communism as an ideological and geopolitical force (the breaching of the Berlin Wall, November 1989) progressive shifts of economic (and political) power to Asia, the development of the Internet and mobile telephony, the realisation that climate change is a global threat, the arrival of the Anthropocene era. The crash of Lehman Brothers (September 2009) that presaged the global economic crisis and events such as the Greek sovereign debt crisis, the 'Arab Spring' and the emergence of ISIS and global terrorism have challenged any notion of a stable world. The result has been manifested in mass migrations of refugee populations, the emergence of new political alliances and the rise of influence of the far right in Europe and America. At the urban scale economic liberalisation, transnational conglomerates, global flows of capital, new and 'smart' technologies and the splintering of political allegiances have fundamentally altered the role of city government. Constant change and uncertainty now prevail, and there are few signs that this will diminish in the immediate future (Fig. 4.1).

Under 'liquid modernity' our lives, cities and institutions cannot keep their shape for long. Change is a constant, and has become a 'permanent condition of human life'. The global passage from solid to liquid modernity has confronted us with a radical set of challenges. Social forms and institutions no longer have the time to solidify and act as potential models from which to extrapolate strategies for an increasingly uncertain future. New strategies are required for the individual and the organisation. These involve a tactical approach – splicing together a series of short-term projects that are steadily replacing formal long-term plans. 'Liquid modernity' is characterised by a form of nomadism – strategies of constant adaptation. The

Fig. 4.1 Uncertain times. (image: Victoria Lee)

individual changes careers, countries, friends and acquaintances as circumstances prevail. What is trustworthy and works today, may be obsolete by tomorrow. Tactics are formulated, tested, utilised and abandoned as opportunities open up and close down. Success depends on a high degree of adaptability and resilience.

4.3 The Challenges for City Governance

Like individuals, successful cities are responding with adaptive strategies where agility and resilience are seen as the new hallmarks of stability. City government is redefining itself as strategist, enabler and facilitator rather than sole provider. The politics of persuasion are replacing the politics of direction.

This poses a significant structural challenge because city government is rarely agile. City government originated as a way of providing a collective defence against external threats and over time evolved into new institutions designed to protect the rights of the citizen, initially as property owner, and then as individual. The nineteenth century industrial city required more complex forms of city government that could respond to the challenges of public health, public safety and transportation. Then, in the twentieth century, city government became the direct provider of universal services such as utilities, schooling and parks. It became a major landowner, had the right to levy taxes, became a provider of housing and developed delivery systems for social care and welfare. In other words, it became a bureaucracy, sometimes efficient, sometimes unwieldy. As city government grew, it necessarily became specialised and organised into professional departments.

While bureaucracies may at times be efficient deliverers of services they are seldom suited to dealing with the so-called 'wicked' issues that require interdisciplinary working. This weakness became increasingly problematic in the final decades of the twentieth century. The new issues facing cities were less tangible and discrete and far more interconnected. Economic restructuring, crime and social marginalisation, disparities in health, energy and climate change, and demographic change began to dominate the agenda. Ageing and ethnically diverse populations posed far more complex challenges around social cohesion. In the UK, the essential roles of city government came into question. With the rise of the neo-liberal economy, restrictions in public funding generated interest in alternative forms of partnership, procurement and service delivery (frequently involving contractual partnerships with the private and community sectors). Local authorities moved from being service providers to strategic enablers and commissioning agencies. In some countries the fallout from economic crises has led to the significant weakening of government at the city level. In Athens for example, the debt crisis has forced citizens to set up autonomous local organisations of self-reliance to fill the gaps caused by the collapse of basic health and welfare services.

There are four further factors impacting on the effectiveness of traditional city governance. First, where politicians are elected on a time-limited term, the electoral cycle plays a significant role as a dampener on innovation and risk-taking. With low electoral turnout, in the UK typically 28–35%, a supposedly 'safe' seat can rest on a majority of less than a hundred votes and can easily be overturned by unresolved local issues. Consequently significant strategic matters, upon which the city's future prosperity and wellbeing might depend, can easily become side lined if they run beyond the next electoral cycle. It is easy to accuse city government (especially at the local level) of being parochial but that is precisely what it is meant to be.

The second factor at play is the role of data. Cities are a long way behind private corporations in seizing the importance of 'big data' and have been particularly slow in responding to new forms of social media and communication. These are already altering the political process and in North America and Europe, for example, mastery of social media has already caused major political upheavals. The beneficial or cynical utilisation of social media and citizen data for urban planning is still in its infancy and the implications for the design community are still relatively unexplored.

Third, it is a mistake to believe that there is always a single optimum solution to a problem and that consensus can be achieved through the application of a particular set of techniques. Belief in this leads to the dangerous territory of marketing and consumer (citizen) manipulation. Design at the neighbourhood or city level is ultimately about making choices between competing options, and these choices will have ramifications for the individual citizen. As discussed later in this chapter these citizens are increasingly demanding a say. Urban planning and urban design are deeply embedded in the politics of the city – they are inherently political actions.

Finally, the technical nature of the planning system has now increased to the point where it is difficult for the public to understand, let alone participate in its workings. In the UK, Local Plans tend to be long, technical and often dull policy documents, of little direct relevance to the communities that they are meant to serve. As a result local planning authorities have often lost the engagement of their citizens. The fragmentation of political consensus has led to the emergence of an important 'third voice' in the process. Well-organised communities are playing a key role in the city planning debate often from outside the formal political structures. At one end of the spectrum are NIMBY's (Not in My Back Yard) simply seeking to protect their interests; at the other end are articulate and skilful individuals and lobbying organisations. In addition throughout the world a new constituency is emerging among disenfranchised youth who no longer trust or respect their leaders and wish to have a political voice of their own.

4.4 Rethinking Planning

Translated into the fields of city planning and urban design these factors are causing cities to rethink how they plan for the future and the mechanisms through which they can implement change. Where city government is democratically elected, traditional methods of top-down planning are no longer tolerated. There is a growing expectation for debate on future plans and development. Proposals put forward by technical experts or city leaders are unlikely to be accepted without explanation and the right to comment.

Where power is concentrated in the hands of an individual, or a small elite, top-down, autocratic, strategies may still be deployed in the form of grand projets driven by a formal master plan. There is a long (and at times distinguished) history of individuals – often popes, emperors and dictators – imposing their particular visions on cities. These projects are often celebrated today although their impact at the time, in terms of the misery for the urban poor, is largely forgotten. The tradition has continued to the present era, with examples such as Ceausescu's destruction of swathes of central Bucharest.

There are also examples where urban planning can take place largely in isolation from a political context. One-off events such as Olympics and Expos, the establishment of new capitals (Canberra or Brasilia) and in the UK the post war New Towns, established under the 1946 New Towns Act (Harlow or Stevenage) are all examples where conventional masterplans and design strategies were used.

Notwithstanding the circumstances where more formal master plans might be deployed, the standard condition that prevails in the post-industrial city is a process of incremental expansion, densification and constant renewal. Whether urban change is being driven by public or private sector interventions there is an increasing expectation of a wider democratic debate about the possible outcomes. This is the condition that exists in European, Australasian, North and South American cities. A wider discussion on the objectives and form of urban planning is also emerging in many Asian and Russian cities, where the expectation of active citizenship is a side effect of the connected global economy. The pressing everyday challenge, therefore, is to devise new mechanisms for planning existing cities that are under a constant process of renewal. In this respect, the question is more about how it might be possible to shape, rather than plan cities.

It is a common misconception that urban areas owe their existence to the individual genius or villain, the planner, the architect or the developer. This might have been the case historically (and indeed may apply under conditions of dictatorship or single party rule today) but in a democracy power is diffused and is exercised by multiple stakeholders (Fig. 4.2). These include landowners, government bodies, politicians, technical experts, community and interest groups, individuals and development companies. Planners, architects and their clients have to be able to navigate an increasingly complex environment. The diagram below (Fig. 4.3) was produced by Design for London (the Mayor of London's Architecture Studio, 2006–2013) to illustrate the number of different stakeholders that might be involved in the design and construction of one piece of new open space in London. It is not meant to explain how the city works, rather it is meant to show its complexity. Complex as the city is, occasionally, if one can find a path through this labyrinth and put the right people together and understand how different agendas can be reconciled, change can happen. And sometimes this change can be good. City planning is emerging into an agency that might broker agreements and build alliances between different actors and stakeholders. Under these conditions new institutions and skill sets are required. The successful planner or urban designer is in fact a navigator and needs to work within the mechanics of the city: its processes, politics and dynamics.

Fig. 4.2 Design charette. (image: Design for London/Peter Bishop)

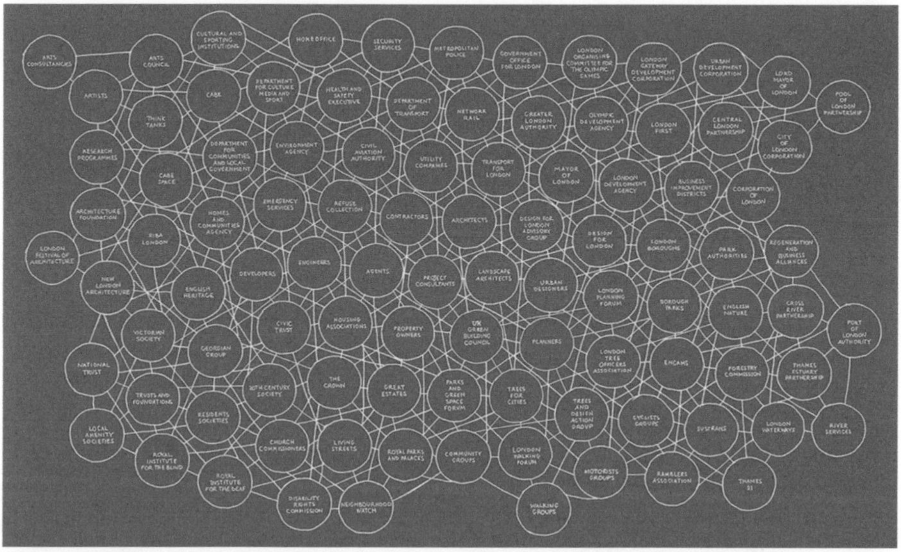

Fig. 4.3 The number of stakeholders involved in designing one open space in London. (image: Design for London/Peter Bishop)

4.5 Four Dimensional Design

We know that the city is perpetually adapting and evolving. The evidence lies all around us where the historic layers of cities co-exist in a rich mosaic of contrasting architectural styles. Sometimes these historic structures are embedded successfully within the life and function of the modern city, sometimes their outline is written in its present street patterns and sometimes they seem stranded incongruously in its fabric. This four-dimensional city is the reality in which we work. Yet much urban thinking and many strategies are still strictly three-dimensional. City authorities are continually seeking permanent and final solutions for an elusive 'end state'. In a 'liquid' world strategic planning processes are increasingly unsuited to the pace of modern urban change, plans are often out-dated before they are even published, while on a day-to-day basis the control of development perpetuates categories of use that are inflexible and unsuited to times of continuous change.

Following the 2009 financial crisis many city authorities in Europe and North America that are charged with the task of encouraging the revitalisation and redevelopment of their urban areas are now finding that for the most part they lack the resources and political power to control and implement formal planning strategies. As a result, some are beginning to experiment with looser planning visions and design frameworks. When linked to packages of smaller, often temporary initiatives, they can unlock urban change now, rather than at some future date. Such approaches are finding resonance and support in the some of the emerging new multi-disciplinary architectural studios in the UK and elsewhere. A more incremental

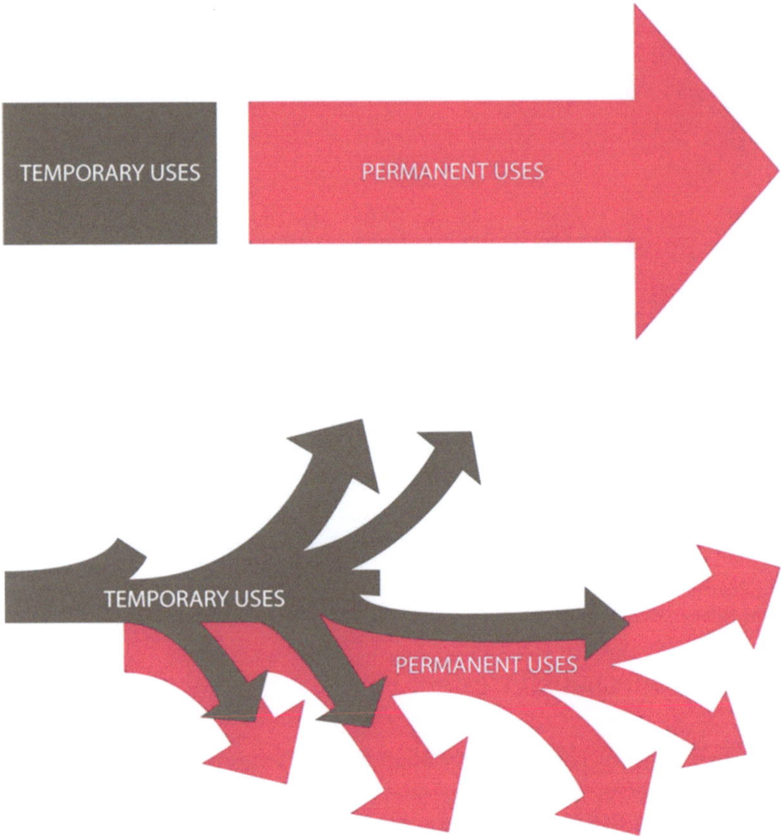

Fig. 4.4 Temporary urbanism, a process not a goal. (Image: London Development Agency/Peter Bishop)

approach is also eliciting an increasingly sympathetic response from landowners and developers. Institutions are now recognising that their plans need to be more flexible and phased over longer time periods. One outcome is an emerging interest in temporary activities or interim phases of development.

The emergence of temporary urbanism (Fig. 4.4) has resulted in an upsurge of "pop-up" or "meanwhile" projects – shops, sports facilities, business premises and artists studios, restaurants, cinemas, bars and theatres (Bishop and Williams 2012). Some are clearly making use of a glut of vacant property as developers realise that they can reduce risks through short-term leases. Short-term activities offer security and an income, but more importantly they can demonstrate the future potential of the place or building (Fig. 4.5). There is also a cachet associated with time-limited exclusivity. This has both novelty value and consumer appeal.

Fig. 4.5 The temporary city. (image: Studio Egret West/Peter Bishop)

Until recently temporary activities had attracted little professional interest from architects or planners. Yet, from the travelling fairground to the parking lot through to the designer pop-up restaurant such activities have always been a vital part of urban culture. They fill the gaps and enliven the urban experience, and they can bring considerable benefit when sensitively incorporated into urban planning. Strategies that utilise temporary activities recognise the essential transience of urban life and are likely to be more effective in an uncertain age (Bishop and Williams 2012).

Many of these "bottom-up" interventions or fleeting re-configurations of space are seemingly spontaneous or arise without consent. Andy Beckett, writing in The Guardian notes:

> in the crevices the developers have left behind, there is a counter-trend at work. You can see it in the guerrilla gardening movement and the boom in music festivals; in the vogue for temporary "pop-up" shops, restaurants and cinemas in empty urban spaces; in the artists occupying disused high-street stores from Durham to Margate; in the sudden appearance and popularity in London of outdoor ping pong tables; and in the Edinburgh crowds last summer queuing to see spooky late-night art installations in the city's usually staid Royal Botanic garden. (Beckett 2010).

The new wave of temporary activities may just be a passing fashion that will fade away once the novelty value has diminished. But there is also a possibility that it represents a more fundamental shift in the use of land and buildings with deeper implications for urban policy, design and practice.

4.6 Tactical Urbanism

Where the conditions of market certainty, financial and legislative control do not exist together masterplanning must take a more tactical approach. De Certeau defines 'strategy as the work of systematizing, and imposing order. Its ways are set, and it cannot break up and regroup easily, something that a tactical model does naturally. Tactics are the ways we negotiate the strategies that are set us, or the creative intelligence that we employ to ensure our environment works to meet our needs. Tactics allow us to assert our individuality and autonomy. Tacticians rely on timing and are always on the watch for opportunities that must be seized 'on the wing'' (de Certeau 1988).

The notion of a more tactical and pragmatic form of urbanism can be traced back to the writing of Jane Jacobs who had a significant influence on the 'everyday urbanism' of Margaret Crawford, John Chase and John Kaliski (Chawford 2008), and 'tactical' or alternative masterplanning by individuals such as Stephen Marshall (2008). The everyday urbanism proposed by Crawford is concerned with interventions that reinforce the heterogeneous qualities of small, temporary, not-intended, undistinguished, though well-used spaces. It takes ordinary places, 'the nooks and crannies of existing urban environments', thinks about them in new ways, and makes small changes that may accumulate to have a transformative effect on the wider locality. It aims to reconnect urban design with ordinary human and social meanings and thus strengthen 'the connective tissue that binds daily lives together'. Kirshenblatt-Gimblett (2008) refer to the tactical as the 'vernacular' – what ordinary people do in their everyday lives that is often outside (or in spite of) planning regulations. And because the vernacular is tactical its practices have a temporal character arising and melting away at short notice (Fig. 4.6).

A well known example of grand masterplans going wrong is Christopher Wren's proposals to rebuild the City of London after the Great Fire of 1666. Wren's ambitious proposal was to re-plan the chaotic street plan of the medieval city into a rational, logical city of the enlightenment with grand avenues, squares and public buildings. King Charles II was enthusiastic but Wren had made a fundamental mistake – neither he nor the King had any control over the land, and there was no way of financing his proposals. The freeholders who owned the many small parcels of land were already rebuilding on their existing plots (defined by the medieval streets), whilst the ink was drying on Wren's plans (Schofield 1984).

Wren would have been advised to have looked at an earlier example of incremental city planning, in Rome the century before. In the sixteenth century Rome was a medieval slum. The Papacy was effectively powerless and bankrupt but wished to transform the city into a fitting stage for the centre of Christendom. Pope Sixtus V (1580–85) commissioned the architect Domenico Fontana to produce a new masterplan for the city. The resulting plan situated Rome within its wider landscape with grand avenues aligned with distant viewpoints and new piazzas and civic spaces. It was ambitious and could not possibly have been implemented due to the papacy's lack of power and resources. In recognition of these constraints work on the plan

Fig. 4.6 Union Street London, one of seventy small projects in the Bankside Urban Forest project by Witherford Watson Mann. (image: Peter Bishop)

was commenced by symbolically placing six obelisks and columns salvaged from the ruins of Ancient Rome at key nodal points on the plan. Gradually spaces were cleared around them to form piazzas and these were then connected. So strong was his vision that Fontana's plan of new straight avenues, public squares and fountains was implemented by subsequent rulers. The process took over 200 years but Rome, as they say, was not built in a day.

It is extraordinary that many masterplanners today ignore the fundamentals of land ownership, political control and finance. Without control of these elements, masterplans are doomed to failure. As Sixtus V and Fontana demonstrated – if power and control are lacking, get tactical.

4.7 Experiments in City Government and Flexible Strategies

Some cities are now experimenting with new organisational frameworks to deliver incremental strategies involving new alliances between political leaders, health and social providers, the private sector, academia and professional bodies. This is leading to a growing emphasis on project implementation, rather than abstract planning and it has obvious political advantages. Tangible benefits can be delivered in the immediate short term, while also establishing momentum towards longer-term goals. It is of course possible that the longer-term vision may not be achieved (with the attendant risks of disillusionment and cynicism). However, in an age of uncertainty this is often seen as a risk worth taking. What is significant is that urban planners are starting to recognise that incremental strategies based on specific projects can offer a powerful means of mobilising public interest. Embodied in this approach is an enthusiasm for experimentation and a willingness to start a project, see what happens along the way, and adapt as necessary. This represents a powerful mechanism to re-tune our cities for whatever lies ahead in an uncertain and 'liquid' world.

These changes of approach are being reflected in architectural practice. Following the ground-breaking work of Urban Catalyst in Berlin, small multi-disciplinary architectural practices (involving architects, graphic designers, artists etc.) are increasingly experimenting with new types of urban projects (Urban Catalyst 2007). These are often highly entrepreneurial and seek out spatial opportunities in the city. This generation of practitioners will develop their own projects rather than wait to be commissioned. These projects may be temporary uses on a small piece of urban space, may be polemical and challenge the orthodoxy of established urban land allocations, or celebratory and explore the potentials of urban spaces for new forms of activity and enjoyment (Fig. 4.7).

The challenge is to devise new government structures that might nurture and harness these non-traditional incremental and tactical design techniques. One of the upshots in the UK is the experiment with Mayoral government. The rationale for this new breed of mayors in the UK is the need for strategic leadership and problem

Fig. 4.7 Berlin, February morning and –4 celsius. Tactical urbanism, small projects make a difference. (image: Peter Bishop)

solving. City Mayors are specifically viewed as agents of change and social and economic renewal. This trend is illustrated by the election of the architect George Fergusson as Mayor of Bristol (2012–2016). Ferguson, an independent, stepped outside the party politics and repeated formulas that had dominated city political life and was elected on a specific manifesto to improve the environment and fabric of the city.

Elsewhere in Europe, the work in Tirana (Albania) of Mayor Edi Rama has been particularly influential (Farago 2016). Elected in 2000 this artist and writer commenced the renewal of his city using radical and imaginative (and cheap) initiatives such as the repainting of building facades and public space improvements. In 2004 he was awarded the prize of the World's best Mayor, the citation stating, "Edi Rama is the man who changed a whole city. Now there is a new Tirana, coloured, happy, with a new and improved infrastructure and public life." One of the drivers for civic renewal in Albania is Atelier Albania, a collective of architects and designers and external advisors. It has set out an agenda based on three principles; pilot projects based on international; redefinition of the brief of the National Planning Bureau through a dialogue of stakeholders; the development of a coherent national design strategy, 'The Metabolism of Albania (Atelier Albania 2016).

An earlier and much cited example of alternative approaches to city strategy making can be seen in the work of Jaime Lerner in Curitiba, Brazil (Mayor 1971–75, 1979–84 and 1989–93, (now professor of urban and regional planning at the Universida de Federal do Parana). Lerner was an architect by background and when elected Mayor was a political outsider. Through necessity therefore he by-passed the formal departments of his city administration and devised a wide range of innovative city strategies ranging from public transport to public space and waste management. All were centred on the daily needs of the citizen, particularly the urban poor. Lerner linked up many city problems into innovative multi-win strategies. He incentivised school children to collect and recycle garbage in return for food left over from the city market each day. He 'bought' rubbish collected by fisherman as an alternative 'catch'. In response to citizen requests for a metro (wholly beyond the financial limitations of the city at the time) Lerner designed a 'metro at street level'. A set of dedicated bus priority lanes (innovative at the time) provided fast and reliable connections across the city. The bus stops were designed as metro stations with ticket barriers and platforms for fast boarding. The result was a cheap, fast reliable and high status transport system for the city. Lerner transformed his city without the requirement for formal city departments and in the absence of major public funding.

Barcelona, under Mayor Pasquale Maragall (Mayor 1982–1997), bid for the Olympics in order to rebrand itself and renew its essential infrastructure. There was a formal Olympics masterplan to create a sports campus within the city, but Barcelona was the first (and along with London in 2012) the only example of a city using the Olympics tactically to set a new course of urban regeneration. The strategy was to reconnect the city with the Mediterranean and to transform its public spaces. Critically architectural design and public space were placed at the heart of the Mayor's strategy (Monclus 2011). In the run up to the 1992 Olympics, the Mayor established a panel of design advisors and advocates and they proposed significant investment in the streets and public spaces of the city. The 1992 Games was a catalyst for urban change and this change was design led capturing the international imagination. It was not the Olympics alone that led to the regeneration of Barcelona, but the associated investment in public space.

4.8 Design for London

The work of Mayor Maragall in Barcelona had a significant influence on Richard Rogers (one of Barcelona's design advisers) in his work as Chair of the Urban Task Force (HMSO 1999) This thinking was further developed when Rogers was asked to be design advisor for London under Ken Livingstone, Mayor between 2000 and 2008. In 2000, Livingstone and Rogers set up a small strategic design think-tank called the Architecture and Urbanism Unit. This was enlarged in 2006 into Design for London with an explicitly experimental remit on urban planning, strategy making and design. Its brief from the Mayor was to 'think about London, what makes London the place it is and to devise strategies for making London better'. A typical example is the East London Green Grid, commenced in 2006 and incorporated into the London Plan in 2012. This was a 25-year project to connect green space into a network that reconnected east London with the surrounding countryside.

Livingstone had already articulated the vision for London as a 'global city' competing on a global stage, open, tolerant, business-friendly and with a social and environmental conscience. Regeneration had to go hand in hand with sharing the benefits equitably throughout the population. Design for London was given licence to respond to this political vision and devise spatial design strategies for its realisation. The operational approach was summed up by the phrase 'catch and steer'. The purpose was to shape, rather than plan the city by 'catching' projects and 'steering' them in different directions. It worked in conjunction with a wide range of partners and collaborated with architects, planners, researchers, and local authorities. The design philosophy of the studio was distinctly one of tactical urbanism. It worked by producing (Bishop 2013):

- Big strategies: single powerful ideas that could then be turned into incremental development programmes through alliances with partner organisations.
- Small change: projects that explored ways in which urban change could be promulgated without direct control of long-term funding or powers of regulation.
- Policy making: taking strategic control of the procurement of architects to work on London Government projects and writing policy papers on housing and area regeneration.
- Campaigning: running a series of exhibitions to engage Londoners in a debate about their city.

Over a 6-year period, and with an aggregate total budget of under £20 million, it developed policy documents on housing design standards (GLA 2011), public space (GLA 2010a, b), procurement (GLA/TfL 2012) and environmental standards (London Plan 2016). It worked with partners to develop and implement:

- 16 new public spaces in London;
- Three large-scale spatial strategies (the Royal Docks, East London Green Grid (Figs. 4.8, 4.9, 4.10, and 4.11) and the Olympic Fringe Master plans);
- Three comprehensive town centre regeneration programmes (Dalston, Barking and Deptford).

In addition, Design for London was at the forefront of developing new methodologies for local regeneration and participative design (Fig. 4.12; MuF/LDA 2009). Temporary urbanism was part of its methodology to prototype ideas, stimulate regeneration, involve local communities and place making (Bishop and Williams 2012).

Design for London had a deliberately ambiguous position within the structure of London government. It was cross-agency and reported through a board to the Mayor. While it had no formal powers and only limited funding, it was perceived to have political clout and this gave it influence beyond its size. It operated by identifying areas of potential urban change and then fashioned bespoke intervention strategies. Its effectiveness relied on its ability to influence events and to partner different organisations. It had the tacit support of the Mayor, although no one really knew the extent of its political influence. In fact, this ambiguity provided its soft power and

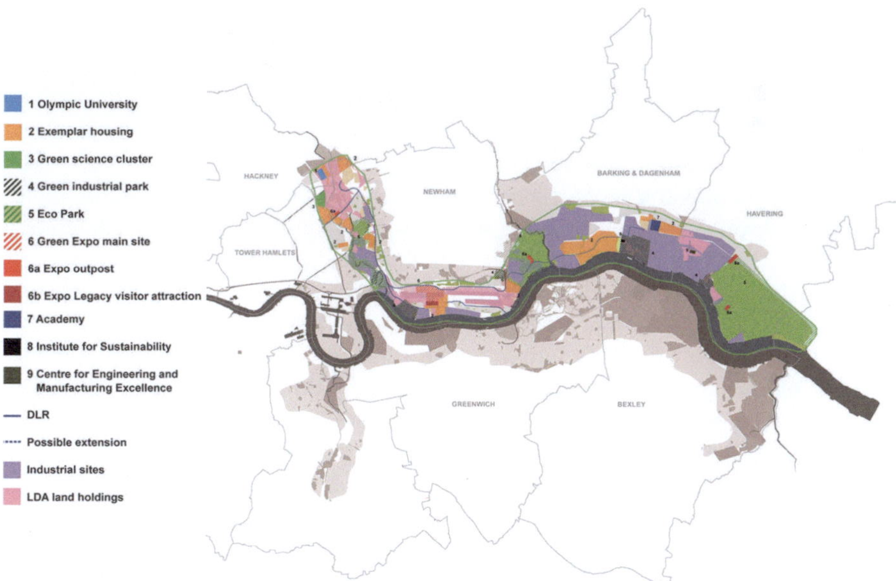

Fig. 4.8 East London Green Enterprise Zone. (image: London Development Agency/Peter Bishop)

Fig. 4.9 East London Green Grid, big ideas – small projects. (image: Design for London/Peter Bishop)

influence. The result was a unique body of work that has had a disproportional impact on the development of the city.

Design for London's approach to urbanism was to develop big ideas and implement them through a myriad of small steps. Its design philosophy and methodology was routed deeply in English pragmatism rather than large-scale 'grand projects'. It was also heavily influenced by the European traditions of urban planning and design, especially the work of practitioners in the Netherlands and Germany (Bishop 2013). It deliberately eschewed traditional forms of influence through the planning process but sought to exert tangential influence through conceptual thinking (Fig. 4.13). In essence, it was seeking to embrace Bauman's concept of 'liquid modernity' (Bauman 2004). Kees Christiaanse described this as 'a negotiated approach, an urbanism of brokerage' (Professor of Urban Design ETH Zurich and member of DfL's Design Advisory Panel 2006–2009). Kieran Long (Editor of the Architects Journal 2007–2009) described its work as 'using guerrilla tactics to become the most influential city architects office in the country' (Architects Journal 2008). Its overarching objective was to harness London's growth to create a better city for all its citizens.

Fig. 4.10 East London Green Grid, big ideas – strategy. (image: Design for London/Peter Bishop)

4.9 The Implications for Design Education

There is no single accepted definition of Urban Design. The term was first used specifically at a conference at Harvard in 1956 (Harvard Urban Design Conference 1956), although earlier work carried out by a number of eminent architects would today be clearly recognised as urban design.

Rather than being a discipline in its own right (with a canonical body of professional knowledge), it represents a particular perspective and approach that may be brought into being through the "joint work of the architect, landscape architect and urban planner". There are many definitions of Urban Design but a common element is a recognition that it encompasses different disciplines in order to enquire into the form and function of the city as a place of human coexistence or conflict. Urban design, by its very nature, has an open-ended time frame. It should also be able to operate simultaneously at different scales from the street block to local neighbourhood, the city and the region. Inevitably therefore urban design is concerned with multiple contexts, including the spatial, the social, the economic and the environmental. Finally the design of urban areas must recognise its impact on the condition of diverse populations. It has multiple clients and therefore any intervention will

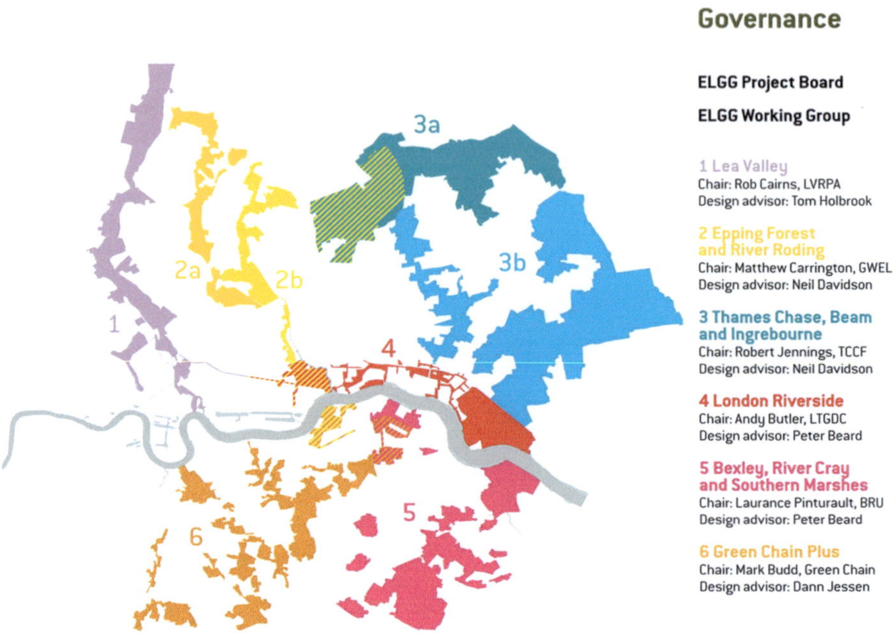

Fig. 4.11 East London Green Grid, big ideas – governance. (image: Design for London/Peter Bishop)

Fig. 4.12 Incremental strategies. (image: Muf Architects/Design for London)

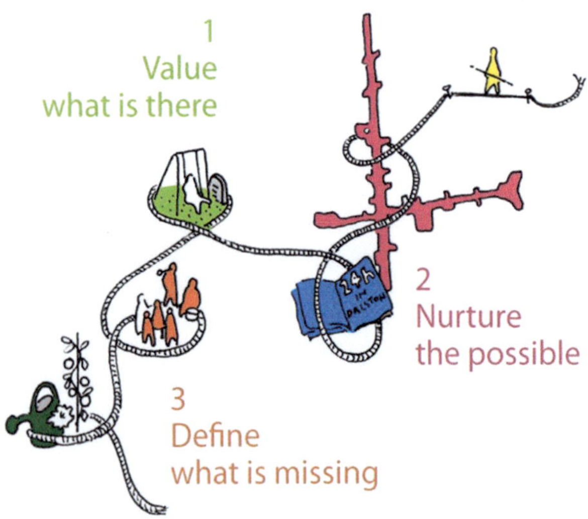

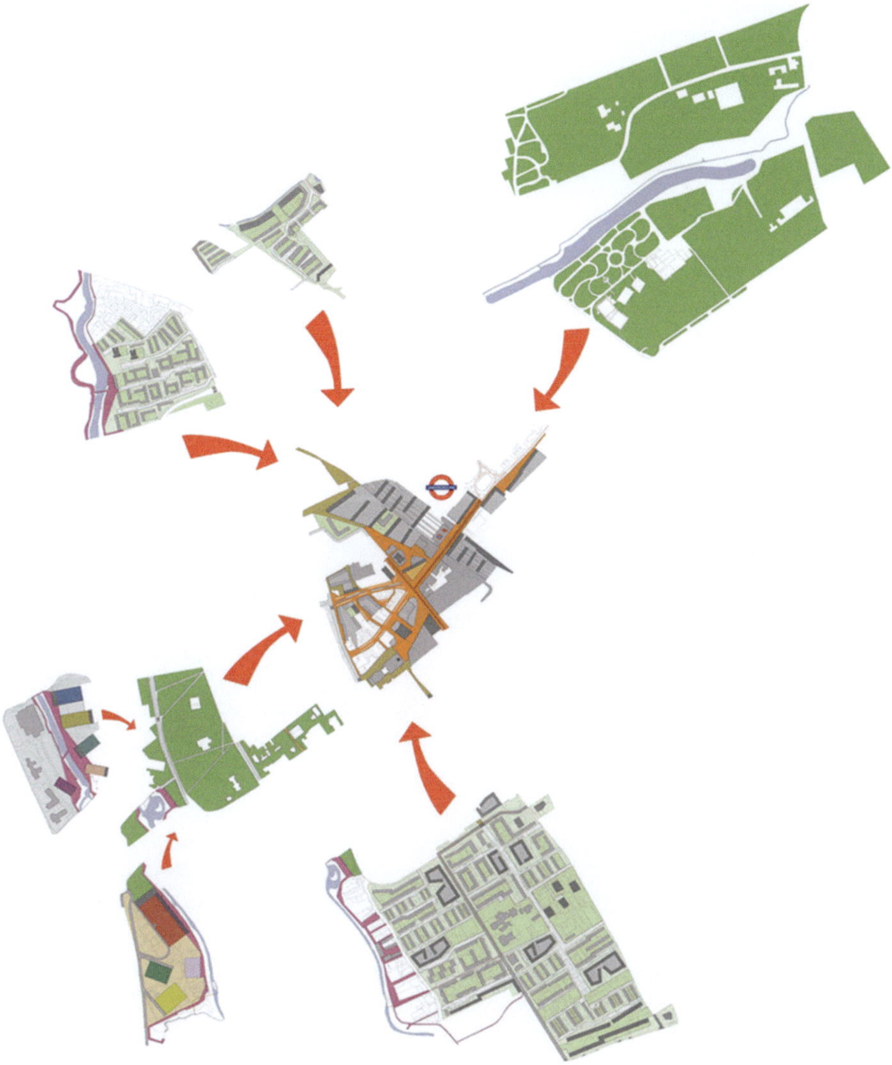

Fig. 4.13 Tactical urbanism, taking the city apart and putting it back together again. (image: Design for London/Peter Bishop)

benefit some, possibly at the expense of others. Interventions at the city scale do, therefore, have political implications.

The challenges of cities in the present 'liquid' age raises some interesting questions for design academics, educators and practitioners alike. Given the overwhelming evidence that cities are a complex overlay of buildings and activities that are in one way or another temporary, why are designers still so focused on permanence? Why is there so little consideration given to what will occur when the particular building,

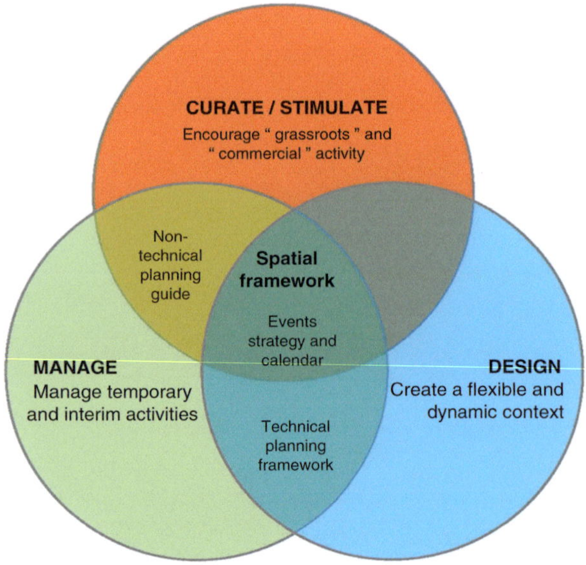

Fig. 4.14 Temporary urbanism, a process not a goal. (image: London Development Agency/Peter Bishop)

space or city form becomes redundant and will need to be replaced in the future? Are the present day drivers of urban change themselves transient or might they represent a more enduring influence on the form of cities? Could temporary uses be a manifestation of the emergence of a more dynamic, flexible or adaptive urbanism, where the city is becoming more responsive to new needs, demands and preferences of its users? And if so, do the systems of regulation and planning need to adjust to the requirements and implications of this new fluidity without stifling creativity (Fig. 4.14)?

These questions will need to be answered if we are to avoid costly consequences in the future. The concept of sustainable development will need to be translated into real and tangible design solutions if our towns and cities are to avoid serious future costs. Increasingly the design debate will move towards innovation in the design of housing, workplaces, public spaces and transport (Fig. 4.15). This places real challenges on the ways in which the next generation of practitioners will have to be trained and educated.

Design education can play a pivotal role in encouraging innovation and disseminating lessons and experiences. This calls for architectural education to expand its role beyond pure 'bricks and mortar' to become a centre for (policy) debate and best practice that embraces the wider design, development and construction industry.

4.10 City Government and Universities

Across the world new relationships are developing between universities and their host cities. Internationally, UN-Habitat Hubs has set up consortia of universities to collaborate on thematic priorities, in particular linking universities in the global north with the global south (unhabitat.org). The themes are cross-cutting and geared

Fig. 4.15 Curation of public space, Gillette Square Dalston. (image: Design for London)

towards the common critical issues facing cities. For example University College London is the lead on Safe Cities, while the University of Naples leads on Urban Regeneration. Central to the initiative is that universities will become partners with city government, actively engaged in real problem solving and closing the gap between academia and practice. This restructuring of education and research is reflected in new courses and research clusters. Stanford, for example, has established a Human Cities Initiative specifically aimed at producing the next generation of city leaders. Again, the chosen themes are deliberately cross-cutting including sustainable cities, smart cities and public space. Another initiative by the University of Tampere (Finland) – Urban Education Live (UEL) – aims to create and test a new model of collaboration between universities and urban communities. In this model, universities would act as catalysts for urban change through trans-educational urban capacity building. UTS is setting up a Sydney Urban_Institute, which aims to implement innovative solutions for contemporary urban problems in collaboration with private, public and people organisations.

In the UK, the Higher Education Funding Council (HEFC) is placing a growing emphasis on research impact. In response, UCL has set up a set of new thematic research clusters including the Centre for London Urban Research and Design (CLOUD). This will provide a new focus for research, debate and projects that address spatial design issues in London. The core work of the Centre will focus on how innovative architectural and urban design can address the spatial and environmental design issues facing London.

These, and many other initiatives, reflect the fact that all cities are struggling with very similar issues; problems around housing, escalating land values, transport investment, climate impacts and governance. There are three key drivers behind the

Fig. 4.16 London as a collage of land uses. (image: Design for London)

need for greater collaboration between city government and universities. First, city governments often lack the ability to assemble and evaluate long-term data about the city's built environment or to carry out independent research and devise strategies on architecture and urban design. Second, new policy frameworks are required that embrace complex spatial and environmental issues. Third, cities need independent forums for enquiry and debate. As these bridges develop between government, practitioners and academics there are growing opportunities for the global exchange of new ideas and approaches as well as collaborative research.

4.11 Conclusions

In today's fluid world the major challenges facing cities are encouraging a move from formal design and planning initiatives towards dynamic open-ended strategies for urban choreography and space management (Fig. 4.16). They are also leading to a reappraisal of traditional approaches to architecture and urban design. Simplistic urban forms will not give rise to rich and diverse cities that will be sustainable centres of innovation and civic life. An over planned and over regulated city is a sterile city. The response will in the future lie less in the physical design of space but in the processes through which it is created, managed and adapted. These same challenges are resulting in new opportunities for academic engagement on teaching and research as city government seeks new alliances with universities and practice.

References

Atelier Albania (2016) The metabolism of Albania: activating the potential of Albanian territory. iabr/UP

Beckett A (2010) In the gaps developers left, another world is being built. The Guardian, Saturday 21 August 2010, p 30

Bishop P (2013) Design for London – Bartlett research portfolios. University College, London

Bishop P, Williams L (2012) The temporary city. Routledge, Oxford

Crawford M (2008) The current state of everyday urbanism. In: Chase JL, Crawford M, Kaliski J (eds) Everyday urbanism, 2nd edn. The Monacelli Press, New York

de Certeau N (1988) The practice of everyday life. University of California Press, Berkeley/Los Angeles

Farago J (2016) The Guardian, November 2016

Greater London Authority (2010a) London housing design guide

Greater London Authority (2010b) London's great outdoors

Greater London Authority (2011) London Plan

Greater London Authority (2012) GLA/TfL architecture, design and urbanism panel user guide

Greater London Authority (2016) The London plan 2004, 2020, 2016. Greater London Authority, London

Harvard Urban Design Conference (1956) Urban design – progressive architecture. August

HMSO (1999) Towards an urban renaissance. Subsequent urban white paper, 2000

Kirshenblatt-Gimblett B (2008) Performing the city: reflections on the urban vernacular. In: Chase JL, Crawford M, Kaliski J (eds) Everyday urbanism, 2nd edn. The Monacelli Press, New York

Long K (2008) Editorial. Architects Journal, April 2008

Marshall S (2008) Cities, design and evolution. Routledge, Oxford

Monclus F-J (2011) Barcelona 1992. In: Gold JR, Gold MM (eds) Olympic cities: city agendas, planning and the world's games, 1896–2016. Routledge, New York

Muf Architects and JL Gibbons (2009) Making space for Dalston, LDA

Schofield J (1984) The building of London. British Museum Publications, London

Tester K (2004) The social thought of Zygmund Bauman. Palgrave McMillan

Urban Catalyst (2007) Urban pioneers: temporary use and urban development in Berlin. Jovis, Berlin

Chapter 5
Redirecting the City?

Gerard Reinmuth

Abstract In framing the inaugural Seoul Biennale of Architecture and Urbanism, Co-Curators Hyungmin Pai and Alejandro Zaera-Polo suggest "the cities of the world stand at a crossroads." Calling for a new urban cosmology within which to think through the distribution of the "emerging commons" they suggest are imperative to fairly organise and distribute to avoid inequality and environmental decay. This call comes from an acute awareness of the incapacity of current political and economic systems to address fundamental risks to the planet such as climate change and ever-increasing inequity among its inhabitants. Debate rages between the assertion that engaging in current political and economic frameworks can only result in the reproduction of the inequities upon which they are based, and the alternate view that perhaps these current structures might be re-appropriated to different ends.

The discipline of architecture can surely be called upon as a key instrument in this project, but for two conceptual barriers – a resolution on the matter of engagement with existing structures, and a rethinking of the discipline that might trigger new professional formations more suitable to this task than the profession of architecture as it is currently constituted. I will explore, by example, a specific apparatus that attempts to navigate the disciplinary and governmental impasse that sits before us. The context for thinking through the opportunities for and limitations of this apparatus is foregrounded by bodies of scholarship across two separate but interrelated themes: the replacement of politics by management (Ranciere, Zizek, Morton etc); and a consideration of the professional anxieties of the discipline of architecture as it pertains to this political condition (Harvey, Cunningham, Swyendouw, Lahiji, Deamer et al.).

G. Reinmuth (✉)
Faculty of Design Architecture Building, University of Technology Sydney, Broadway, NSW, Australia
e-mail: gerard.reinmuth@uts.edu.au

R. Roggema (ed.), *Contemporary Urban Design Thinking*, Contemporary Urban Design Thinking, https://doi.org/10.1007/978-3-319-91950-8_5

5.1 Introduction

In framing the inaugural Seoul Biennale of Architecture and Urbanism, Co-Curators Hyungmin Pai and Alejandro Zaera-Polo suggest "the cities of the world stand at a crossroads." Calling for a new urban cosmology within which to think through the distribution of the "emerging commons" they suggest are imperative to fairly organise and distribute to avoid inequality and environmental decay. In accepting Aureli's duality (Aureli 2008) that a political project is a spatial one and a spatial project a political one, it follows that a new cosmology and cosmopolitics (Zaera-Polo 2016) cannot be bought into existence without significant rethinking of access to and sharing of the commons, fundamental to which is their spatial organisation and distribution.

The call for new cosmological/cosmopolitical frameworks comes from an acute awareness of the incapacity of current political and economic systems to address fundamental risks to the planet such as climate change and ever-increasing inequity among its inhabitants. The discipline of architecture can surely be called upon as a key instrument in this project, but for two conceptual barriers. Firstly, a rethinking of the discipline is required which must then trigger new professional formations more suitable to this task than the profession of architecture as it is currently constituted. Simultaneously, a context within which redirection of the city and its commons must be found.

This second point is a matter of some complexity and debate. Erik Swyngedouw (2016) suggests that, regardless of the professional formation, architecture can only spatialise, reproduce and distribute current inequities, unless it engages in a "set of affective and sequential acts that. . . inaugurate an equal, solidarity-based and free socio-spatial order that abolishes what exists." His view is based upon the assessment that the design of current managerial practices to ensure the evacuation of the political from the work of the spatial disciplines. Challenging this, Harvey (2010) cautions that one should be careful not to demolish the current structures that capitalism has produced when the possibility remains that this machinery might be re-appropriated to different ends (Cunningham 2014).

I will suggest that the concept of autonomy – understood in architecture as pertaining to the autonomy of an object from external forces, and the idea of a subsequent archipelago of "autonomous objects" as having the potential to resist external political or economic forces, should be rethought. I propose instead that the conception of the object in architecture as the after-effect of relations (Benjamin 2016) gives us a mechanism whereby we might reconsider the autonomy of the object itself. If we were to replace "object" with "city", we start to consider the reorganization of relations at a larger scale where the redirection of existing systems might counter the current political hegemony where flows of capital have been excised from the political as a locus of contestation.

This reformulation leads to the search for an appropriate means of engagement for such a project. I will explore, by example, a specific apparatus that attempts to navigate the disciplinary and governmental impasse that sits before us. The context for thinking through the opportunities for and limitations of this apparatus is foregrounded by bodies of scholarship across two separate but interrelated themes: the replacement of politics by management (Ranciere, Zizek, Morton etc); and a consideration of the professional anxieties of the discipline of architecture as it pertains to this political condition (Harvey, Cunningham, Swyendouw, Lahiji, Deamer et al.).

5.2 Post Politics, the Domination of flows and the Individual Versus the Multitude

> "The ultimate sign of 'post-politics' in all western countries [is] the growth of a managerial approach to government: government is reconceived as a managerial function, deprived of its properly political dimension." (Žižek 2002)

Slavoj Žižek (2009), Jacques Ranciere (2007) and others have proclaimed this the era of the post-political, a condition where the political subject is de-subjectivised by the onslaught of an exclusionary market and the absence of any real choices for those within it. The post-political, according to Zizek, exists now that "the political sphere is discarded as a sphere of mobilization and politics is reduced to a social-pragmatic exercise in implementing and managing developments that are regarded as inevitable (BAVO 2007)".

Another way of understanding this situation is as a political condition where capital is excised from the locus of contestation and debate such that all possible alternatives assume at their centre that the market rules. Choices or options exist on the basis that there are none that will fundamentally challenge or contest the prevailing conditions. New waves of privatization and deregulation meet little opposition, property speculation continues to restructure access to the city and its "commons" and (as starkly described by Thomas Piketty (2014) intergenerational wealth now re-emerges as the most significant determinant of an individual's future opportunity.

The complexity of politics as an ideological contest has been replaced by the rhetoric of self-organization and emergent systems that work to minimize resistance to the continuity and optimisation of flows. The management of flows – of capital and of those with capital – is the key preoccupation of contemporary governance practices, mirroring in the core ambitions of successful businesses to subvert all decisions to the improvement of efficiency, lowering of costs and increase in profit." (LeCavalier 2016). Jesse LeCavalier has described this in spatial terms as having seeded a dominant culture of logistics, a culture that strips back understandings of national borders, trade across these border and the labour laws negotiating modes of production within them as nothing more than obstacles to ambition" (LeCavalier 2016).

Assisting in this reshaping of this political and economic landscape has been the rise of a specific language that magnifies the critical distance between political decisions and their impacts, presenting a world where the compromises in these decisions are explained away through a series of questionable claims such as the "cohesive city", or "sustainable practice" (Morton 2013). These part-paradoxes exist not only in marketing but underpin significant and well-respected developments in architecture, landscape and urban design. For example, a new frontline in the negotiation of spatial disciplines, ecology and the city is known as “ecological urbanism” – a concept described by Lahiji in fairly extreme terms as based on “the preposterous claim that the most effective way to attenuate climate change is through large-scale, privatized, suburban developments.” While many aligned to the ecological turn in urbanism practices may reasonably contest such a brutal critique, the presence of such an impossibly large blind spot exists among many of those undertaking so called “sustainable” practices – that is, the management and regulation of capital flows have similarly been embedded in sustainability practices, where sustainability is understood primarily as the sustainment of the current order (Morton 2013).

Locating this discourse more centrally in the language of the architectural discipline for a moment, we can talk specifically of parametricism as conceived by Patrik Schumaker (2008) of Zaha Hadid Architects. The advent of parametricism formally signalled a new direction in Hadid’s practice – from a designer of buildings as specific, place-based interventions based in a concretization of imaginary flow lines, to one where everything gives way to the relentlessness of free-flowing uninterrupted surfaces. As the practice expanded from the design of individual objects to urban scale tasks, these uninterrupted volumes and surfaces came to describe entire city precincts. The city, for all its complexity and layering, came to be rendered as object – a conceptual position that amplifies fundamental shortfalls in the profession of architecture, outlined later in this chapter. Not only does the object now rule, but the formal gymnastics underpinning the later incarnation of this practice are located in space of increasing abstraction, as opposed to specific places – a conceptual posture that Timothy Morton argues is impossible to maintain given the way in which global warming has reasserted that place most certainly matters. “We humans,” he warns, find ourselves in this era of global “on a very specific planet with a specific biosphere (Morton 2013).” Parametricists dismiss any consideration of place beyond its existence as an abstract field with no logic outside that which can be measured but insist that the discipline should resist the temptation to make any spatial intervention contrary to the desires of market forces as architecture’s accommodation to the existing social order must be absolute: “it is not architecture’s societal function to actively promote or initiate political agendas that are not already thriving in the political arena” (Schumacher 1997).

In the design of cities as uninterrupted flow paths we have reached a moment where the tsunami of capital appears unstoppable as it finds the most advantageous conditions for replication. Yet Cunningham has pointed out that this reliance on maintaining and optimizing flows to perpetuate the current order may be an Achilles heel where the dependence of our urban environments on these flows may be “one of the most vulnerable human arrangements that has ever existed (Cunningham 2010).” Like Cunningham, I question how one might interrupt this “perpetuum mobile” in the interests of greater equity.

5.3 Scale and Governance

In any project working to reconfigure and broaden opportunities in the urban realm, the fundamental negotiation is between the individual and the multitude. It is therefore not surprising that in understanding the disadvantage that an individual can face in this paradigm of unstoppable flows, renewed interest has emerged in how to effect resistance or redirection. A key variable here is the scale at which relative ideological independence may be exercised. For example, small-scale groups readily form in protest against large-scale development, loss of heritage, major transport projects and so on and become elevated to political organizations. Yet, the single-issue focus of these communities eventually reveals a lack of universalizing aspirations and ultimately of instrumentality in a broader sense. (Cunningham) This failure to exercise agency beyond specific protests is considered an integral part of the post-political paradigm, where such "acting out" of protest is accounted for and even invited as a desubjectification tool. Ultimately, the inability to reach beyond specific concerns and mobilise a larger spectrum has led to Zizek, Swyngedouw and others characterising these groups as a hysterical acting-out of politics that is "not truly political because of the restricted nature of the constituency (Marchart 2007)."

The current impasse around climate change is an obvious example, where individuals can make symbolic acts "whether at the level of personal recycling or corporate-led "carbon trading," but, as Morton (2013), Lahiji (2014) and others have noted, a full coordinated response is beyond the reach of any collective political body existing or imagined. The argument is made that even if it were possible to imagine exactly what form such a body might take concerns immediately appear regarding "what forms of "authoritarian" power it might have to possess (Cunningham 2014)." This hesitancy leads us to the broader question of what conceptual transformation in the configuration and form of the state might take us beyond the post-political condition. As Harvey bluntly frames this dilemma, while revolution is quite properly opposed to prevailing notions of the republic of property, the presumption that the world's six and a half billion people can be fed, warmed, clothed, housed and cleaned without any hierarchical form of governance and outside the reach of monetization and markets is dubious in the extreme (Harvey 2010).

This impasse – between the highly visible but low impact protest group, and our failure at planetary scale to make even marginal adjustments to account for planetary-scale problems – can in part explain the rise of the city as the poster child for urban and societal transformation. Concerns regarding the capacity of government to effectively allocate, create or redirect resources in the age of neo-liberal managerialism –led to a focus on city-scale governance as perhaps the most viable arena for consequential action. We have seen this in operation recently in the United States, as the Trump presidency has found resistance from city administrators and Mayors who suggest they will openly resist his administration's policies in regard to climate change or other matters.

Sassen and others have argued that the emergence of some "new geography" of global processes could, in itself, provide an opening for the articulation of new forms of "citizenship," not least on the part of architectural practice, whereby the "de-nationalising of urban space and the formation of new claims by international actors, raises anew the perennial question, 'Whose city is it?' (Sassen 2007). The need to reform the entire political and management apparatus is based on the premise that "spatialised policies (planning, architecture, urban policies, etc.) are among the core dispositifs of such post-politicizing governing (Swyngedouw 2016) and the key mechanisms through which fundamental conflicts are avoided. At present, when political conflicts are revealed, they are dismissed as a form of extremism and not in the spirit of appropriate "dialogue" and "deliberation" organised around the logic that no real choices, nor real opportunities for proposing serious alternatives, actually exists? (Cunningham 2014).

The question remains as to whether, in a post-political-reality where consensus has been built around the inevitability of state-backed capitalism as an economic system architecture can "state its own claim and interrupt the police order?" (Lahiji 2014) Cunningham cautions that, in considering a reform in the way architects work and the "relative potential of larger scale strategic work and the inevitable intersection of this work with government," (Cunningham 2016) we cannot succeed without the political strategies at stake within it being extended "to the entire complex of institutions, and not only those involving architecture and building" (Cunningham 2014). Libero suggests that such an intervention should embrace "legal, political, and financial" matters and the attention to "cooperative arrangements like community development corporations and land trusts" (Deamer 2014).

Defining the city as the realm of re-politicization requires us to be more specific about distinguishing between its physical and nonphysical parameters. The physical environment is the structural basis, the three-dimensional frame of the socio-political sphere. However, analysing the political potential in a dense urban context places the emphasis not on the production of architecture, but rather on the production of space (Lahiji 2014). The question then, is whether autonomy at the scale of city might provide a breakthrough in where and how we challenge contemporary conditions.

5.4 Professional Anxieties

At the risk of a significant jump-cut in the argument, I am going to suggest that the impossibility of a global compact regarding equity of resources and collective responsibility for issues such as climate change – and a parallel questioning in the architectural discipline about agency – might come together in a consideration of how we understand the appropriate context and apparatus that may contribute to urban transformation.

We have seen how the desubjectification of the individual that defines this post-political age has impacted in disciplinary terms (with ecological urbanism example). At the same time, a trajectory of fragmentation in the profession over the last two decades was accelerated by the 2008 Global Financial Crisis and the anxieties it highlighted in regard to the economic sustainability of practice models. The ever-diminishing role and stature of the architectural profession – in both the processes that underpin building production (being replaced in traditional roles by numerous new consultant types and minimal participation in large-scale decision-making (planning, urban design and strategy) led to concerns about the relevance of the architectural profession to society.

Thus, the post-political condition is that which binds the internal questioning within the discipline (relevance) and the seeming impossibility of the profession regaining any agency. Peggy Deamer (2015) and others have suggested that the only response it to decouple the business model which underpins the profession of architecture from the object, given the impossibility that she sees it that anyone in practice "might think past a client-driven practice and put their spatial expertise toward thwarting private development" (Deamer 2015). Deamer's position is thus not motivated by a disinterest in form and materiality but as a means of breaking the object-based contract the profession has made with capital: "We get published by the object, we are motivated by the object, we staff up and organise our offices around the client's objects" (Deamer 2015).

Thus a transformation of the profession to enable better conditions for the engagement of the discipline with capital is both imperative and yet remains unlikely until the profession's self image can be shifted – from a focus and measurement of success based on the production of objects to spatial engagement at a scale larger than the individual building. Linking the internal and external worlds of the profession is the client-driven nature of professional practice as a service industry, a condition that requires a different engagement with capital. Externally, the question remains of what form of the state (or organising body) might take and how any new form of urban governance is structured and deployed.

Conflating both the professional and larger societal questions, Ross Exo Adams (2014) asks how we make a stand for architecture, to "have a political dimension or role and if so, to what extent, and how might this be understood in the context of the disciplinary and professional formations?" Or, in reverse, in what way is architecture relevant to the discourse of the political? If we accept the generalisation that society is now built on inequality and "politics is normally called by the name 'management' to sustain and perpetuate the same order" (Lahiji 2014) then we must make a stand for the discipline of architecture by making a stand against the current formulation of the profession of architecture.

We are left with two questions: first, where and how might impact be possible on the future organisation of the world and the sharing of commons within it and, secondly, how might the spatial disciplines and principally architecture participate in the reorganisation of the city to effect more equitable outcomes driven by a logic other than the market? I suggest that a shifting of horizons is required to overcome dilemmas inherent in the formation and focus of the profession in contemporary society, a shift that would ideally dovetail into the larger societal and political questions with which the chapter started.

The field of potential responses can be bracketed by two polarisations: the insurgent architect and the idea of autonomy. Firstly to the idea of the insurgent architect, a concept that has gained currency through the work of Harvey, Swygendouw, Schneider/Till and others as they attempt to formulate the idea of a resistant or emancipatory posture via which the discipline might exert political influence. Insurgency has – through their work– been given tangible form as one who works through apparently radical or resistant acts, interventions, exhibitions and other "bottom-up" actions and events that start to loosen current understandings of the discipline and profession. However, this acting out of resistance is predicated upon clear boundaries within which it can occur, thus mirroring the fragmentation of the collective into single-issue pseudo-political groups that also have clear operational boundaries. For these reasons, the critique of Williams and Srnicek to groups such as Occupy could equally apply here – a sort of architectural incarnation of "folk politics" (Snricek and Williams 2015) which lacks the agency required to have any impact beyond that of the immediate event. The authors argue that the fetishisation of immediate results leads to an empty pragmatism that struggles to maintain the present balance of power, rather than seeking to change structural conditions.

While it can be acknowledged that moments of insurgency may have value as part of a constellation of redirective measures, it is argued that such acts, by themselves, are "incapable of articulating or building a new world," (Snricek and Williams 2015) and simply form a sort of smokescreen that enables the neoliberal project: "these practices of performative reconfiguration/displacement ultimately support what they intend to subvert, since the very fields of such 'transgressions' are already taken into account, even engendered by the hegemonic form" (Žižek 1999). In the discipline specifically, Aureli echoes this with the assertion that "the activist and participatory practices that are so popular today are the latest iteration of a reformist syndrome whose pathology is to preserve social and political conditions as they are" (Aureli 2013)." In short, the focus of the spatial discipline (architecture) on the sites and scales of least agency has resulted in the solidification of the view that the discipline is incapable of effecting meaningful change and is doomed to operate as a secondary instrument deploying the logic of an all-prevailing market.

Aureli's critique of this theoretical dead-end has led to his focus on the concept of "autonomy". Aureli's thesis, staked out in The Project of Autonomy, Politics and Architecture within and against Capitalism, makes a claim for the autonomy of the architectural discipline through an analysis of the Italian Autonomia movement of the 1960s and 1970s. Aureli's claim for autonomy is based on a rejection of centralised planning in favour of an approach for the city based in the idea of the autonomous object or archipelago – an object that in its formal organization might somehow encapsulate political agendas. A series of internal contradictions and conceptual simplifications reveal the impossibility of this position and its inward focus which brings the risk, as Tafuri famously suggested, that architecture simply works through the confirmation of its "sublime uselessness" caught in the confusion induced by "an utter alienation mistaken for independence" (Tafuri 1980). Libero Andreotti, writing in response to the question of whether architecture can be an emancipatory project, even suggests that the recourse to autonomy as a conceptual position is a “redundant pleonasm designed to uselessly prolong the debate” (Andreotti and Lahiji 2016).

The potential for agency is thus self limited, for in not accepting the impossibility of current conditions, we find solace in polarisations that might by definition have very limited agency and, at worst, can be understood as an integral part of the desubjectification machine so integral to the post-political condition. A more centrist approach has permeated recent writing on the need to redefine the profession, through, for example, refection on the lack of commitment to planning since the late nineteenth century and the suggestion that, by extension, a new relevance may come from reigniting this engagement. David Cunningham is one of a growing chorus who emphasise the necessity for the discipline to focus on a different scale, suggesting that on a planet housing seven billion people, “forms of mediation, abstraction and impersonality are not only ineliminable but necessary to the construction of new social relations and modes of collective transformation of our increasingly urbanised world” (Cunningham 2014). In short, we need “to rethink architecture’s relationship to wider issues of planning” (Cunningham 2014).

A conceptual wormhole through this terrain is provided by philosopher Andrew Benjamin and his work on relationality. Benjamin suggests that buildings should not be considered individually but rather through their incorporation within a network of relations: “The consequence of such a description is that it then allows any one object – the building as object – to be an after-effect of the relations that pertain within a given conjunction. In other words, any singularity is always an after-effect of a network of relations. Transformation therefore is the process of a coming-into-relation (Benjamin 2016).

The “object to network” repositioning suggested by Benjamin is not to diminish the importance of the object, but rather relocates it. Thinking through relationality is to understand the discipline’s specific contribution is the consideration of relations in the broadest possible sense but their reorganisation specifically in spatial terms. Acknowledging relational thinking as a core disciplinary expertise would lead to a new understanding of the architect as someone who works simultaneously across scales and where the outcome of the work is not automatically a single built object. These scales might range from that of the particular (the study of specific spatial arrangements as the after effect of relations) an understanding of how this particularity works in place (the impact of these specific occurrences on the context in which they occur, as the after effect or counter measure to existing conditions) and an awareness of how this particularity relates to, resists or reorganises global flows in a specific place. This conceptual reorientation of the discipline prepares the ground for working at organizational problems at a larger scale than the object through a relational architectural practice. Subsequently, the potential exists for an alignment between the logic of city-based redirection and reorganization and a professional re-formation.

On the professional front, the move from the object to relations brings with it a change in the relation to capital from an object-based exchange to a broader services-based model. If we are to accept this expansion of professional modes we might clarify a disciplinary skillset and language that would redirect our focus toward questions of complex spatial relations. One consolidated profession might be replaced by a multitude of actors from the discipline of architecture working on a diverse set of problems.

5.5 Case Study: Relationality and the Spatial Framework

> "In most of critical urban theoretical apparatuses, the political is usually assumed to emerge from what might broadly be called a 'socio-spatial' analysis. Put simply, a critical theory of the 'social' is considered to be the foundational basis from which an emancipatory urban politics can (or will) emerge, both theoretically and practically. It is the socio-spatial condition and the excavation of the procedures of its production that opens up and charts the terrain of political intervention and animates the politicizing subject". (Swyngedouw 2015)

We are presented then with world where a politics without contestation appears resistant to intervention in the system of unimpeded flows upon which it depends. This inability to intervene is compounded by problems of scale, exemplified by our failure to address planetary problems at a planetary scale. Meanwhile, the discipline of architecture – which might see itself as an actor in the spatial reorganisation of the city – is constrained by a professional formation that privileges the design of individual objects. But what if we approached the city as a viable entity in which to intervene and as we did so, thought of the architecture as relational in nature and thus able to inform new conceptions of the architect?

A key question that appears is that of the relation between the architect and the state. Given this, an increasing focus on the concept of the "city architect" is unsurprising. For example, the Seoul Biennale of Architecture and Urbanism (mentioned at the start of the chapter) places its focus on an urban commons alongside the emerging role of the city architect through a series of exhibitions and symposia surrounding the event. Despite this new focus, the role of city architect remains unaddressed in most theory on the city and the imperative in challenging current forms of management bureaucracy so central to the critiques of the post-political.

In this context, the "city architect" in Sydney, Australia is notable given that the role has existed uninterrupted since March 1816, when the convict architect Francis Greenway was appointed. Since then, the cities and towns of New South Wales had all major public institutions designed by the "Government Architect" as the role came to be called. In 2015 the role was shifted from one focussed on the design and delivery of buildings and public spaces, to that of a strategic design advisor that produced frameworks and policies. The focus on the object, which organises the profession, was now replaced by a focus on relational thinking and the potential of that thinking to impact on political processes.

This new approach brought several benefits, the most importantly of which in the context of this argument is the removal of the Government Architect from the risk of conflicts of interest. While inevitably working in the context of the Government of the day, the role promotes the delivery of advice that has no direct relation to exchanges of capital (as cautioned by Deamer) and posses the disciplinary skillset and political leverage to effect change.

The question then turns to how the disciplinary skillset is utilised. A specific apparatus is being developed in Sydney in a unique collaboration between the Government Architect and a core group of professionals. This apparatus has emerged from the development of a design-led methodology for place-based spatial strategies that organize people, resources and space at the scale of the precinct. A key opportunity of the work is the embedding of these strategies and resultant logics into political processes and planning systems through a document known as a Spatial Framework (Fig. 5.1).

The Spatial Framework emerges from the realisation that the development of large city precincts (or smaller areas within a precinct) still relies on the coordination of Government and non-Government contributions, even in, and ironically because of neo-liberal development practices. In most cases, the Government's role and contributions to a precinct must be coordinated prior to the involvement of the private sector if Government is to actively determine the parameters for success, even if in purely financial terms. This requires the coordination of and integration with multiple Government Departments that often have competing or contrary objectives for success. This moment, and this need to coordinate across Government as it organises itself around a project, has been identified as a gap or portal where significant re-directive opportunities exist (Figs. 5.2 and 5.3).

The importance of this gap lies in its opening to a series of readjustments outlined in this chapter and upon which some form of agency in the post-political world depends. Firstly, the project itself is often not completely determined when an opportunity is identified. This creates and opening that allows for a widening of a project's potential in the moment of its clarification. Secondly, the scale of operation of the precinct is large enough to bring together actors from State and City government and their respective aims and objectives, but small enough to avoid the impossibility of large scale consensus building as described earlier. Finally, the terrain to be traversed through this gap or portal has led to a model where the Government Architect works collaboratively in this gap with actors from the private sector – architects who have developed expertise in this field and can act as foil to the Government Architect, as government actor, in the negotiations required to execute the work.

The Spatial Framework has been developed as a tool that works within this gap through a design-led methodology for place-based, spatial strategies that operate at precinct scale. The Spatial Framework does this through a process which effects a synthesis between spatial intelligence and substantial engagement processes across the political and bureaucratic realms. Thus the potential of a Spatial Framework lies in its ability to facilitate the coordination of cross-Departmental input and the implications of this on how resources or commons are coordinated in this context and in a specific site, or place. The model acts through the reorganisation of relations that, through the confluence of a specific project and place, precinct by precinct, start to enact a new cosmopolitics. The political agency of the Spatial Framework exists not

Fig. 5.1 Circular Quay, Sydney. (image: Transport for NSW)

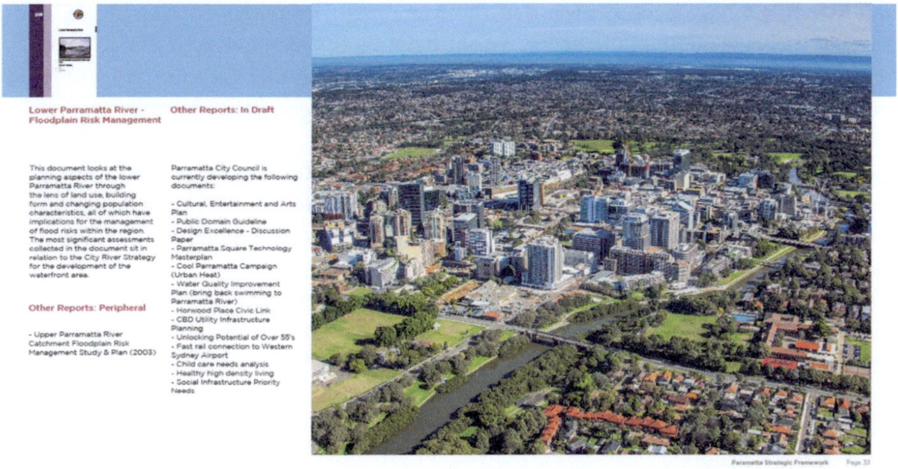

Lower Parramatta River - Floodplain Risk Management

This document looks at the planning aspects of the lower Parramatta River through the lens of land use, building form and changing population characteristics, all of which have implications for the management of flood risks within the region. The most significant assessments collected in the document sit in relation to the City River Strategy for the development of the waterfront area.

Other Reports: Peripheral

- Upper Parramatta River Catchment Floodplain Risk Management Study & Plan (2003)

Other Reports: In Draft

Parramatta City Council is currently developing the following documents:

- Cultural, Entertainment and Arts Plan
- Public Domain Guideline
- Design Excellence - Discussion Paper
- Parramatta Square Technology Masterplan
- Cool Parramatta Campaign (Urban Heat)
- Water Quality Improvement Plan (bring back swimming to Parramatta River)
- Horwood Place Civic Link
- CBD Utility Infrastructure Planning
- Unlocking Potential of Over 55's
- Fast rail connection to Western Sydney Airport
- Child care needs analysis
- Healthy high density living
- Social Infrastructure Priority Needs

Paramatta Strategic Framework Page 33

Fig. 5.2 Parramatta. (image: TBC)

Fig. 5.3 White Bay Power Station. (image: Skyview)

only in its directions, but in the after-effects of the change in political and spatial relations that occur in undergoing the process itself.

The insertion of this tool in gaps that appear at the start of precinct-scale redevelopment means that the tool is not a single protocol but rather talks of a process that in each case emerges into a customised document particular to the question and place where it has been deployed. This focus on customisation within a broadly understood process becomes part of the armament it deploys in convincing each government department "client" as a contingent and specific output that appears to be "ideology free" while at the same time building familiarity with its techniques and processes which promote acceptance of the technique in the bureaucracy.

Key Elements of a Spatial Framework.

Regardless of the specifics of a precinct or question that the Strategic Framework is asked to address, three key elements or platforms are common to the process of creating a framework and modified in detail and emphasis depending on the potential application.

The first of these research and mapping. Given that many of the precincts selected for the Strategic Framework process are important places in the city, there has often been significant research undertaken already and which, despite the specifics of a new project often contain research and analysis that is not only still relevant but which is known to the agencies involved. Thus, the mapping process upon which later analysis occurs has a familiarity which lubricates its political agency but which does not necessarily restrict new propositions. This double-edged sword is exercised through a process of identifying gaps in existing material that in turn provokes fresh analysis. This folding together of existing and new research and insights results in a set of spatial representations, the mode of which is also modified from project to project depending on the actors, the task or the intended audience (Fig. 5.4).

The potential to achieve a level of coordination within and across Government agencies is one of the most powerful aspects of the Strategic Framework process and this forms the second platform. It is also the platform that most effectively utilises the dual-collaboration between Government and external architects, given the nuance of the multiple negotiations necessary to effect. Working on multiple fronts, the Government Architect and their external consultant work in concert at different points in the process, or sometimes simultaneously but on different fronts, to enable specific discussions to occur and/or to achieve specific outcomes (Fig. 5.5).

The process of engagement occurs in multiple layers with key members of Government agencies relevant to decisions required in the precinct and, given the significance of many of the precincts studied, participants tend to be senior bureaucrats with oversight and/or involvement in policy development. Forms of engagement may vary subject to the purpose or context of a Strategic Framework but tend to incorporate up to three phases of engagement. These include: Stakeholder consultation "1 on 1" (a series of meetings with relevant stakeholders that emphasise speaking "off the record" in the interests of gathering intelligence on the various actors); Major Stakeholder Workshops (events that take 1–2 days and engage with representatives from multiple agencies in a collaborative design-led mode where participants are invited to articulating issues, options and possible solutions for a precinct); and Final Stakeholder Review of findings prior to completion of the Framework. The result of this engagement process is a series of Values that identify a collective ambition for each precinct. While single-word values emerge from the initial briefings, workshops and reviews, their detailed articulation is the task of the Government Architect and their external consultants. The precise articulation of

Fig. 5.4 Analysis of circular Quay. (image: Terroir)

Fig. 5.5 Enquiry by design workshop. (images: Michael Ford)

these values, which occurs as the result of a workshop process but elaborated upon after it, enables the opportunity for significant clarification of intent that ultimately informs policy (Figs. 5.6 and 5.7).

Given that Strategic Frameworks precede a potential Precinct Plan, Master Plan or individual projects, it is necessary to articulate the Spatial Principles that the Government has agreed in a clear manner and so this is the third platform. Specific spatial principles are developed that build upon the Mappings and Values such that a clear logic for future precinct development, implicit in which is a position on the "distribution of commons". The Principals and Values are of course related and can work together as an assessment tool to control or guide future development propos-

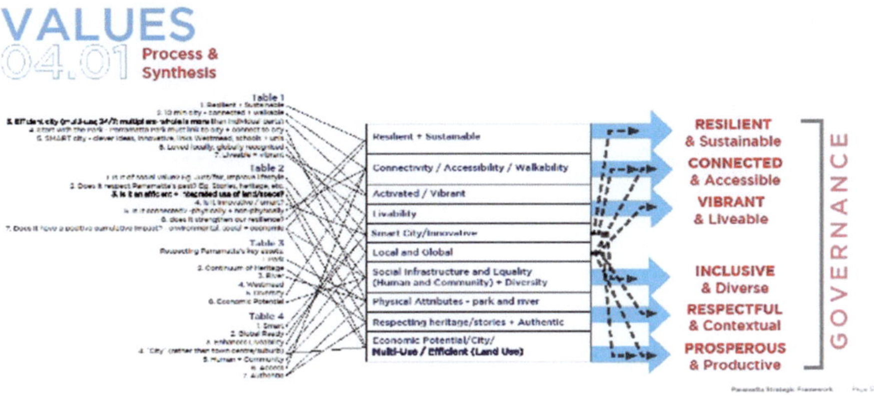

Fig. 5.6 Workshop feedback matrix. (Image: Terroir)

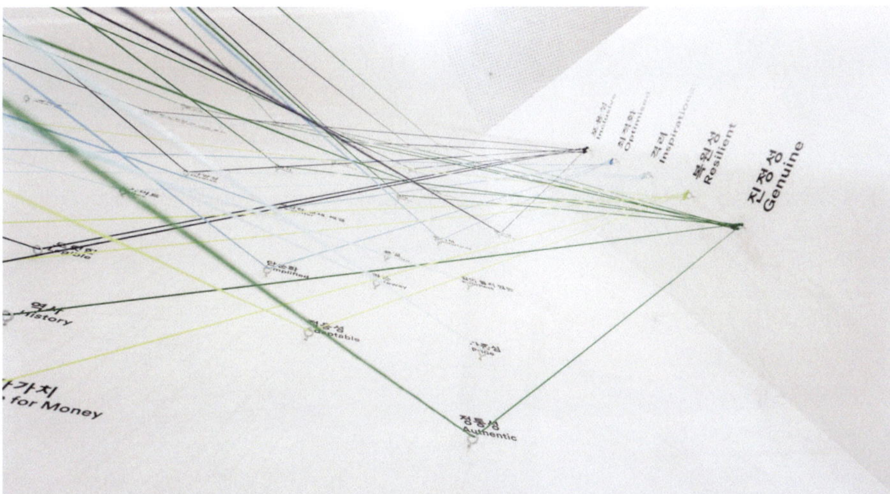

Fig. 5.7 Values, at Seoul Biennale. (Installation: Terroir, Image: Kim Ohrstrom)

als. In some cases, specific assessment tools may also be developed that articulate precisely what it means to comply with the values and principles in any future plans or development (Figs. 5.8, 5.9, 5.10, 5.11 and 5.12).

These three elements work together to form a Strategic Framework document, of which six have been produced in the last 2 years in Sydney and its catchment. In each case, the precinct subjects of each document have been high profile areas subject to imminent transformation on the large scale or renewal or which involve less development but in which important values or characteristics might be threatened. While the fundamentals of the broader political and economic context have not changed, this series of interventions suggests the potential for specific redirections of this paradigm that, over time, may accrete to form a genuine alternative.

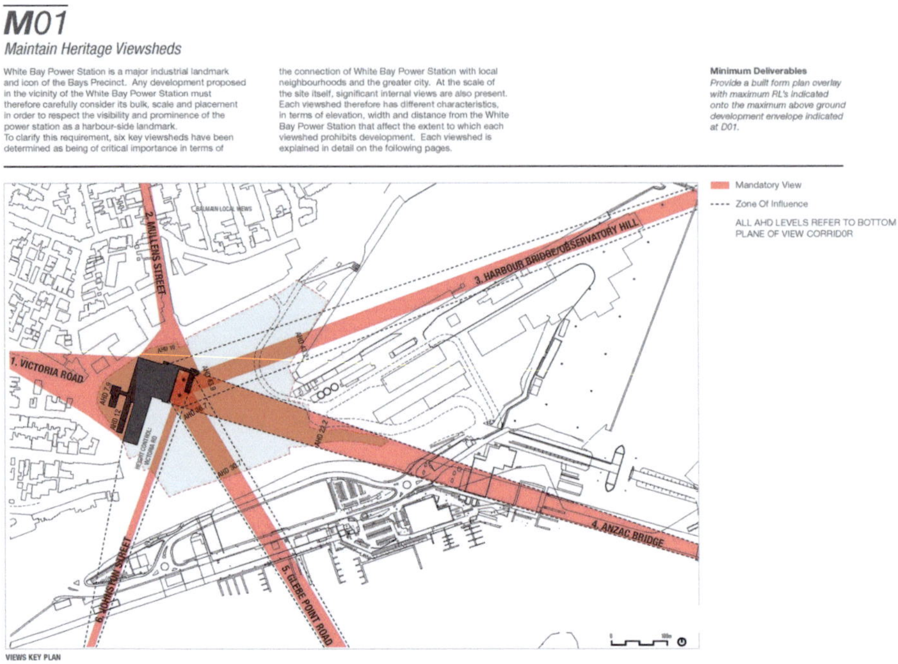

Fig. 5.8 White Bay power station urban design framework, viewsheds diagram. (image: Terroir)

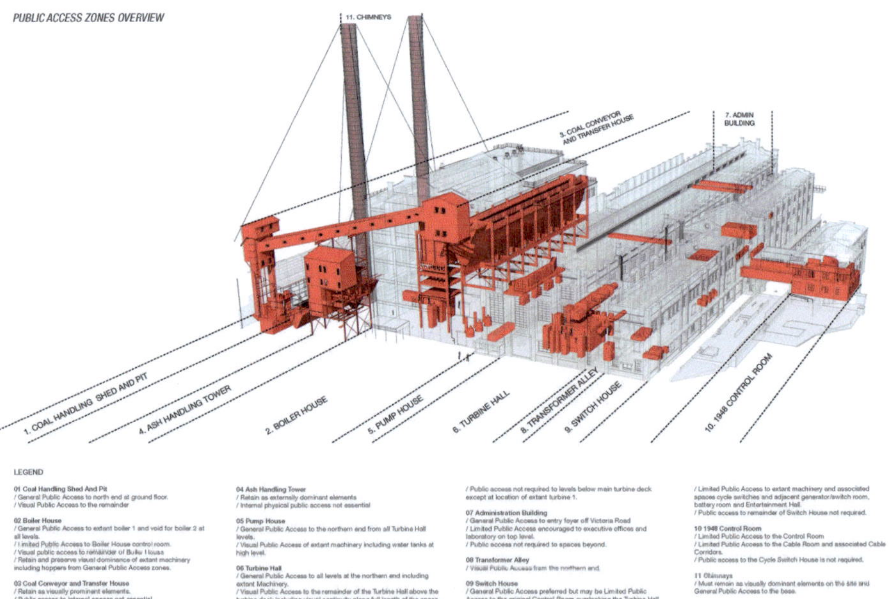

Fig. 5.9 White Bay power station urban design framework, heritage diagram. (image: Terroir)

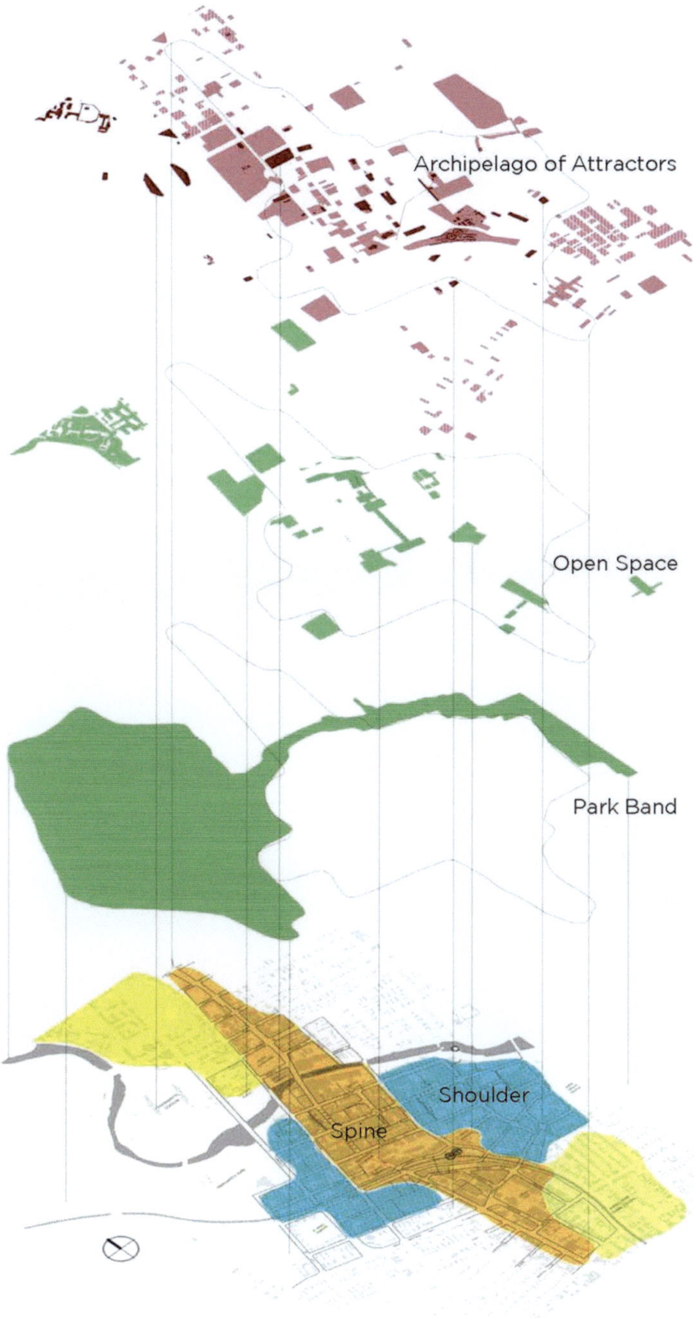

Fig. 5.10 Parramatta strategic framework, principles diagram. (image: Terroir)

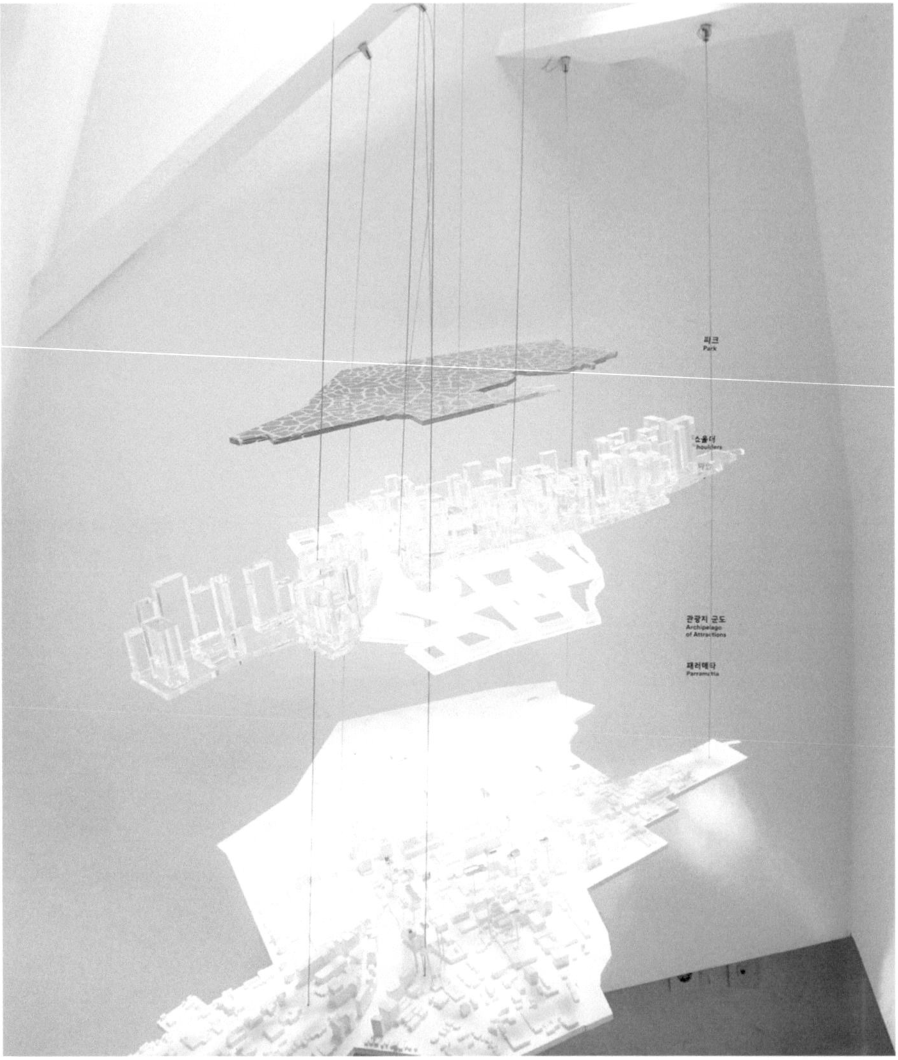

Fig. 5.11 Model of Parramatta strategic framework, principles diagram, exhibited at Seoul Biennale. (Model: Make Models; image: Kim Ohrstrom)

5.6 Conclusion

The argument has persisted in architecture that involvement in the market at any level is to operate without agency, given the all encompassing nature of the neo-liberal hegemony and the impossibility in this post-political era of impacting on it, its insistence on the unchallenged logic of capital and the uninterrupted flows of that capital. This argument still holds almost universally, given the focus of the spatial discipline (architecture) on the production of specific objects that can only reinforce the logic of the market.

Fig. 5.12 Current politics

In examining the multiple scales at which we organise ourself, the city has emerged as the optimum entity at which new forms of governance may form and which can exist with at least partial independence from national or transnational systems that have been demonstrated to have little agency in addressing major concerns. To intervene in these systems, the architectural profession must be rethought as a relational practice as opposed to the object-based practice which defines it currently and which has resulted in engagements with capital that have stripped the profession of agency.

The specific example of the Government Architect in Sydney, New South Wales provides one example of a possible future practice. The Government Architect exploits temporal and organizational gaps in precinct renewal processes and, through the relative independence of the position, can give advice that is free of the conflicts of interest that trouble architects operating through the market. In understanding the potential of architecture as a relational practice, the Government Architect focuses not on specific objects but the creation of platforms, controls and constraints that negotiate the multiple forces of the market, multiple government actors and greater city objectives and strategies. The result is a framework that remains open for private capital and their consultants but where key matters have already been addressed, thus limiting the opportunity for uninhibited development driven only by the logics of the market.

This role of Government Architect is yet to be theorized sufficiently in the debates regarding the future of the discipline of architecture and the profession that acts in its name, while the Strategic Frameworks produced in New South Wales are too recent to be fully evaluated in terms of their impact. However, early signs are promising given a range of decisions made recently in Sydney regarding controls on inappropriate development, allocation of cultural and community facilities and protection of the pubic domain as the direct result of these actors and the documents they produce.

What is possible is that in political and professional conditions, which have been abandoned theoretically as without agency in a post-political world, this example opens up a new front. The logics of governance and city scale, impact on the precinct scale, and a reconfigured understanding of the architect in these conditions (both in terms of their relation to the state but also a focus on architecture itself as a relational practice) all deserve further exploration in the discourse around the development of our cities for a more just redistribution of resources, access, opportunities – the commons – that is so important if we are to insist on a more equitable society.

References

Adams RE (2014) To fill the earth: architecture in a spaceless universe. In: Lahiji N (ed) Architecture against the post-political: essays in reclaiming the critical project. Routledge, London/New York

Andreotti L, Lahiji N (2016) The architecture of phantasmagoria specters of the city. Routledge, London/New York

Aureli PV (2008) The project of autonomy: politics and architectue within and against capitalism. Princeton Architectural Press, New York

Aureli PV (2013) The theology of tabula rasa: Walter Benjamin and Architecture in the age of precarity. The Log 27

BAVO (ed) (2007) Re-imagining democracy in the neo-liberal city. NAi, Rotterdam

Benjamin A (2016) Building, building. In: Araguez J (ed) The building. Lars Muller, Zurich

Cunningham DI (2010) Kraftwerk and the image of modernity. In: Albiez S, Pattie D (eds) Kraftwerk: music non-stop. Continuum, London/New York, pp 44–62

Cunningham DI (2014) Metropolitics, or, architecture and the contemporary left. In: Lahiji N (ed) Architecture against the post-political: essays in reclaiming the critical project. Routledge, New York, pp 11–30

Cunningham DI (2016) Architecture, capitalism and the "Autonomy" of the political. In: Lahiji N (ed) Can architecture be an emancipatory project? Dialogues on architecture and the left. Zero Books, Alresford, pp 87–105

Deamer P (2014) Architecture and capitalism: 1845-present. Routledge, New York

Deamer P (2015) The architect as worker: immaterial labor, the creative class, and the politics of design. Bloomsburg Press, London

Harvey D (2010) Commonwealth: an exchange online at https://antonionegriinenglish.wordpress.com/2010/09/09/commonwealth-an-exchange-between-david-harvey-and-hardt-and-negri/

Lahiji N (ed) (2014) Architecture against the post-political: essays in reclaiming the critical project. Routledge, Oxon

LeCavalier J (2016) The rule of logistics: Walmart and the architecture of fulfillment, University of Minnesota Press, Minneapolis

Marchart O (2007) Post-foundational political thought – political difference in Nancy, Lefort, Badiou and Laclau. University Press, Edinburgh

Morton T (2013) Hyperobjects: philosophy and ecology after the end of the world. University of Minnesota Press, London

Piketty T (2014) Capital in the twenty-first century. Belknap Press, Cambridge, MA

Ranciere J (2007) Hatred of democracy (trans: Corcoran S). Verso, London/New York

Sassen S (2007) Globalisation and the formation of claims

Schumacher P (1997) Materialism versus and morality. Lecture given at the Architectural Association, London. http://www.patrikschumacher.com/Texts/morality2.htm

Schumacher P (2008) A parametric manifesto. Position paper for the dark side Club, Venice Biennale of Architecture

Snricek N, Williams A (2015) Inventing the future: postcapitalism and a world without work. Verso, London

Swyngedouw E (2015) Liquid power: contested hydro-modernities in 20th century Spain. MIT Press, Cambridge, MA

Swyngedouw E (2016) La naturaleza no existe: la sostenibilidad como síntoma de una planificación despolitizada buenos aires: puentoaéro. 94 p
Tafuri M (1980) Histories and theories of architecture. Harper Collins, New York
Zaera-Polo A (2016) The politics of the envelope: a political critique of materialism, vol. 17
Žižek S (1999) The ticklish subject: the absent Centre of Political Ontology. Verso, London
Žižek S (2002) Revolution at the gates: Žižek on Lenin – The 1917 writings. Verso, London
Žižek S (2009) First as tragedy, then as farce. Verso, London/New York

Chapter 6
Microeconomic Reform to Create Activation and Improve Liveability

Justine Kinch

Abstract Sydney will grow by 1.6 million people by 2036. One million of those new people will live in Western Sydney. A region already disadvantaged by a lack of jobs and connectivity. As a society we are increasingly unhealthy: air pollution in western Sydney consistently exceeds acceptable levels. In 1995 56% of Australians were overweight, in 2015 it was 63%… In 2011 18% of the children were obese, in the early 70's it was 4%. And, one in five Australians will experience a mental illness within 12 months. With this ambitious growth agenda and the new three-city vision (Greater Sydney Commission, 2016). The Eastern City, the Central City, and a new Parkland City in Western Sydney, we need to be careful about providing the right support and infrastructure to ensure that equity is restored across metropolitan Sydney. The new Parkland City for Western Sydney, which is proposed to be located in a region with the highest number of inactive adults, highest obesity levels of children in Sydney, poor air quality, poor public transport connections, a jobs deficit of approximately 300,000 and it's about 10 degrees warmer than the Eastern City in the summer. There is a need to address liveability with this planned growth so that the lives of people are enhanced by the benefits of growth. This chapter discusses Micro economic reform and how it can be applied to Australian cities to make them more productive, liveable and sustainable.

6.1 Introduction

Sydney will grow by 1.6 million people by 2036. One million of those new people will live in Western Sydney. A region already disadvantaged by a lack of jobs and connectivity. As a society we are increasingly unhealthy: air pollution in western Sydney consistently exceeds acceptable levels. In 1995 56% of Australians were overweight, in 2015 it was 63%… In 2011 18% of the children were obese, in the early 70's it was 4%. And, one in five Australians will experience a mental illness

J. Kinch (✉)
Arcadis, Sydney, NSW, Australia
e-mail: justine.kinch@arcadis.com

R. Roggema (ed.), *Contemporary Urban Design Thinking*, Contemporary Urban Design Thinking, https://doi.org/10.1007/978-3-319-91950-8_6

within 12 months. With this ambitious growth agenda and the new three-city vision (Greater Sydney Commission, 2016). The Eastern City, the Central City, and a new Parkland City in Western Sydney, we need to be careful about providing the right support and infrastructure to ensure that equity is restored across metropolitan Sydney. The new Parkland City for Western Sydney, which is proposed to be located in a region with the highest number of inactive adults, highest obesity levels of children in Sydney, poor air quality, poor public transport connections, a jobs deficit of approximately 300,000 and it's about 10 degrees warmer than the Eastern City in the summer. There is a need to address liveability with this planned growth so that the lives of people are enhanced by the benefits of growth. This chapter discusses Micro economic reform and how it can be applied to Australian cities to make them more productive, liveable and sustainable.

6.2 Bottom–Up Microeconomic Reform

Microeconomic reform (or often just economic reform) comprises policies directed to achieve improvements in economic efficiency, either by eliminating or reducing distortions in individual sectors of the economy or by reforming economy-wide policies such as tax policy and competition policy with an emphasis on economic efficiency, rather than other goals such as equity or employment growth (Wikipedia undated).

One of the most celebrated bits of micro-economic reform in Australia is the competition policy that the Australian Government brought in during the 1990's (Gittins 2007). And what that policy basically did, was to reform preventing businesses having monopolies. So the idea, that the free market and competition are opened up, because with competition innovation will emerge. And more businesses participating will occur, importantly with the competition policy, occasionally you need monopolies, but that body in the public benefit test for us. This implies that if a monopoly for certain market sector is needed, they had to demonstrate that there were wider community benefits, such as social, environmental, education and health.

Not all microeconomic reforms have obvious or direct economic benefits but the wider benefits that come from improved health, social and environmental outcomes cannot be discounted. The following examples are varied but will hopefully illustrate the many ways in reform can contribute to the overall wellbeing of our community. Below are examples of micro economic reform, bottom up and top down approaches followed by a discussion of the key ingredients for success along the way.

6.3 Playful City USA

One of the favourite programs out of North America is Playful City USA (KaBOOM! undated). This is a recognition program that gets cities to sign up, to say: "we're going to be more playful".

Cities with underprivileged children from low socio-economic areas are targeted. The concepts are typically low tech and cheap but they need sponsorship and leadership, and they need advocacy. As an example, in Evans, Colorado every Sunday in summer a fire truck comes in and sprays water on a field, and the kids run around and play under it, simple.

Another example comes from San Antonio Texas and is called Community Clean up volunteer program (KaBOOM! undated). In this example the community volunteers contributed over 1500 h to keep parks safe and beautiful. Volunteers take the rubbish out, they trim the trees and the bushes, and they make repairs. The benefit is active children and communities working together for a common purpose. This is really important to communities, community gardens it's obvious right? They're great, they reduce waste, could get your hands dirty, they create education, and creativity importantly they bring people together.

6.4 Community Gardens

The wider economic benefits of grassroots ideas such as community gardens (Figs. 6.1, and 6.2) can be felt at the local scale through:

- Provision of fresh produce;
- Reduced waste through composting;
- Education – practical gardening techniques;
- Creativity;
- Bringing people together.

But they don't just happen. The City of Sydney has some 23 registered community gardens but they provide support, guidelines and policies to help communities estab-

Fig. 6.1 Community Garden. (Image: Justine Kinch 2012)

Fig. 6.2 Community Garden. (Image: Justine Kinch 2012)

lish gardens ideas (City of Sydney 2017a, b). Community gardens also align with their wider city vision of Green Global and Connected (City of Sydney 2017a, b).

6.5 Farmers Markets

Once again, a farmers market is a bright idea and can be fantastic for revitalizing town centres and there are examples from all over the world, where a farmer's market has transformed communities. Similar to a community garden it is community focused, but it also provides a platform for the entrepreneur and can attract regional visitation. The wider benefits to other businesses and a towns' prosperity can be improved through initiatives such as these. The benefits of a farmers market include:

- Support sustainable agriculture practices;
- Food and nutrition education;
- Promotion of the consumption of fresh produce;
- Town centre revitalisation;
- Offers consumers more choice;
- Community focus;
- Destination/major attraction.

A good example of this is the small town of Girgarre in Victoria, Australia. The closure of the towns' largest employer saw the loss of 800 jobs. But this tiny town has showed that economic development can be driven from grass roots, new ideas.

One local person suggested that the town starts a farmers market, it grew from six stalls to 150 stalls and has helped to raise funds for the upkeep of sports facilities, local fire brigade, the Returned Services League (RSL) and it funds a car community for medical appointments and it also part funded the local kindergarten until a new provider could be found (Lunsted 2017).

In addition to the market, the town has grown a music culture. The town had no music culture so they grew one! It started with three musicians on a hay bale playing and teaching music, now more than 900 people attend different music workshops. The town now hosts monthly jam sessions and attracts people from all over Northern Victoria. The outcome is that the town is now growing as a result of the wider economic benefits of the Farmers Market and Music Festival.

6.6 The High Line

Synonymous with New York the High Line (Friends of the highline undated) is basically an icon of the city now. Once again, it is also a grassroots movement, where local residents Robert Hammond and Joshua David lobbied to have an old rail line converted into a park for the community. Typically, in large cities, abandoned and degraded infrastructure is associated with areas of cheap housing, cheap rent and shops where you could afford to get everything you need on a daily basis.

Hammond and David formed a group called the Friends of the High Line. They wanted to do something for the community, they wanted to give back, that this would be great for their community to put a park there, that was their idea, a great idea. A few key moves started the process of enabling this project:

- They advocated for change and had the will to drive the project;
- The asset owner was on board and brought in a photographer, Joel Sternfield, who shot some beautiful photos of the High Line revealing for the first time what could be possible;
- Mayor Bloomberg was a supporter of the idea and entered into a shared use agreement with the asset owner. So that's how it started, with really good intentions.

In 2009, it was estimated that the High Line (Fig. 6.3) would return the following benefits to the City (Friends of the highline undated):

Fig. 6.3 New York High Line. (Image: Justine Kinch 2012)

- 400,000 visitors per year;
- Raise tax revenues of $286 million over 20 years.

But the reality was more like (Friends of the highline undated):

- 5 million tourists per year.
- $980 million in tax revenues.
- $2,2 billion in economic activity;
- Accidental Gentrification or Super Gentrification.

For an investor the High Line's stunning success makes a project such as this an easy choice to back. High quality green space draws new businesses and dwellings. There's huge money through redevelopment to be made. So investors will partner with city governments, hungry for a heftier tax base, to do it.

But these obsolete bits of infrastructure generally have people living near them, and often, they are poor, low-income communities of colour, forgotten in the shadows of that very strip of concrete or steel. And, this is true for many of the 17 projects involved in the High Line Network.

So, what has it done for the community? What people really needed were jobs, Hammond says, and a more affordable cost of living. Residents also said they staying away from the High Line (Fig. 6.4) for three main reasons (Friends of the highline undated):

- They didn't feel it was built for them;
- They didn't see people who looked like them using it;
- They didn't like the park's mulch-heavy programming.

Hammond has spoken about the importance of how to keep public parks inclusive, as a bigger challenge than the design solutions. 'Instead of asking what the

Fig. 6.4 New York High Line. (Image: Justine Kinch 2012)

design should look like, I wish we'd asked: What can we do for you?' Hammond and others have now opened up the question of equity. The concept of accidental gentrification is seeing upward pressure on rent and housing costs and the nature of retail and other services changing in the area.

Those findings have led to several new initiatives. In 2012, Friends of the High Line launched a suite of paid jobs-training programs aimed at local teenagers, listening programs, focused on environmental stewardship, arts programming, and educating younger kids (Friends of the highline undated). The organization also started to partner with the Elliot-Chelsea and Fulton Houses, the two public housing projects, to develop their programming schedule (Friends of the highline undated). That's how 'ARRIBA!', a summer series of Latin dance parties got started—a resident thought it up, and it's been a big hit. Friends of the High Line also started putting on occasional events within the public housing campuses themselves, avoiding the swarms of tourists.

Perhaps more critically, Friends of the High Line could have worked harder from the start to advocate for affordable housing. Hammond likes to say that his park gets too much credit and too much blame for Chelsea's explosive makeover—city zoning codes were already changing to support redevelopment in the mid-2000s. But the fact is, the High Line (Fig. 6.5) has become a symbol of the "new" New York, a city of profound inequality. Luxury high-rises and catwalk clothiers have taken the place of Chelsea's old eateries and butchers; the neighbourhood income disparities are among the cities most extreme (Friends of the highline undated).

There is a view that the city could have been pushed harder to keep more of the land values that the High Line created for affordable housing and public services; although a zoning amendment approved in 2005 did encourage some low- and middle-income rate units to be built, it wasn't nearly enough. Hammond agrees. "There could have been more government action through zoning changes".

Fig. 6.5 New York High Line. (Image: Liz Ligon, courtesy of Friends of the High Line)

6.7 Summary of Bottom-Up Microeconomic Reform Initiatives

The key successes from the examples above include:

- Need identified eg: community
- Project Sponsor eg: Lord Mayor
- Partnerships eg: Business, Local government
- Governance: Policy/Guidelines/Leadership
- Funding and Finance
- Management

6.8 The New Western City

In 2016 The Greater Sydney Commission developed the three city vision for Greater Sydney (Greater Sydney Commission 2016). Out of this vision comes a number of different government-led initiatives for Western Sydney to drive economic development, improve liveability and promote sustainable solutions for development.

A key enabler for the development of a new Parkland City is the Government's announcement and funded commitment of $5b to build a new airport at Badgerys Creek for Western Sydney. Around the airport we know that there is going to be an aerotropolis. If you Google aerotropolis you get all sorts of different things. But what we understand from the government is that the new aerotropolis is going to be based on logistics, advanced manufacturing, education, tourism with some residential. So right now i there is no infrastructure, there is no transport, there is no energy, water, there is no waste, there is nothing there yet. It represents a once in a lifetime opportunity to plan and deliver a framework for a new city that meets the needs of the communitytoday and into the future.

In addition to the airport and the aerotropolis there are other government-led initiatives in place that need to be considered and have the potential to drive liveability.

6.9 City Deals

"The City Deal will bring together all three levels of government in a collaborative partnership to realise the potential of Sydney's outer west" (Department of the Prime Minister and Cabinet 2017). The aim of the City Deal for Western Sydney is to support:

- Infrastructure investment;
- Programs of employment;
- Housing affordability;
- Environmental and liveability outcomes
- Coordination between governments to deliver regulatory reforms that better integrate infrastructure, land use, housing and environmental planning decisions to facilitate growth.

Three City Deals have been announced in Australia for Townsville, Western Sydney and Launceston. For Western Sydney the City Deal is going to be leveraged off the airport, and that's great. The idea is that there was a Memorandum of Understanding signed in 2016 between federal and state governments to work with local government representatives to commit to driving economic growth and prosperity for Western Sydney. It's the sort of thing government should be doing anyway but at least this is a genuine commitment to do something for Western Sydney, remembering it is already disadvantaged.

How does this work when there are eight local government areas, each with their own constituents and identified needs? The answer is, that we don't know yet but collaboration across the region is essential to ensure that the existing centres of Penrith, Blacktown, Liverpool and Campbelltown complement each other rather than drain or dominate.

Right now in Western Sydney nobody or one group is really responsible for integrated thinking about how we deliver the essential, enabling infrastructure in an optimised, equitable and resilient way. It is expected that an outcome of the Western Sydney City Deal is the establishment of some sort of development or delivery authority that will mandate collaboration cross the various agencies and service providers. This would be a great outcome.

6.10 Innovation Ecosystems

Vibrant innovation ecosystems that bring together new, high-value businesses, researchers and related service providers are emerging across Australia and around the world. These places are stimulating competition, attracting investors and workers, and generating much-needed economic growth. They will drive the next wave of jobs in Australia, as traditional sectors such as manufacturing and mining contract to a smaller share of the economy. There is a big opportunity for our existing centres of Penrith, Blacktown, Liverpool and Campbelltown to leverage this concept, attract the wealth generators of the future and play a significant role in the growth of Sydney.

Simply building infrastructure, a new university or a business campus won't in itself attract innovators. Instead, key stakeholders must agree on a clear long-term vision, which encourages them to take responsibility in the place-making process. This is consistent with the Brookings Institution definition of an Innovation Ecosystem as 'being a combination of economic, physical and networking assets combined with a supportive, risk taking culture to facilitate idea generation and support commercialisation' (Katz and Wagner 2014).

We also believe there is a need to reassess traditional assumptions about cities, such as the idea they should have single main business centres. If Australia can grow new areas of innovative business activity, then our cities will become more polycentric. Polycentric cities have multiple business centres with schools, parks and other facilities close by, making it more feasible for residents to live closer to their workplace. This can have dramatic benefits, including less time spent commuting and a greater sense of local community.

Key factors are:

- Government to make the first move
  - Land rezoning: the decision to rezone land can be instrumental in the success of driving reform and productivity.
  - Funding: innovation hubs/districts can funnel city revenue back into the local community for renewal
  - Leadership: Strong leadership can create a sense of urgency around economic change, and promote innovation as part of a region's identity
- Strategic anchor to be fostered:

  - Universities: Universities attract like-minded organisations. Universities are at the heart of how maturing cities build their intellectual capital and competitive advantage.
  - Proximity: It's not enough to have anchor tenants located near each other. Places and cities must be designed in such a way to encourage the intermingling of ideas and people.
  - Shared Vision: An effective method of ensuring investments are in sync is to form development partnerships between government, private industry and educators.

- Integrated infrastructure plans

  - Transport: The age of driverless cars may seem a long way of, but forward-thinking innovation ecosystems are already preparing for them.
  - Data Links: It's more than digital plumbing.

- Businesses nurture the local community

  - Culture: Without a vibrant community, a district will not be an attractive place to work and live.
  - Low Income Housing: Affordable housing should also be combined with affordable commercial space to encourage start-ups and incubators programs such as those that exist at the Australian Technology Park (ATP).
  - Education: While universities are at the heart of successful innovation centres, non-tertiary education plays an important role.

6.11 Examples of Successful Innovation Ecosystems

6.11.1 The University of Cambridge UK, a Public Initiative

This is an example of a public body taking the initiative in building an innovation ecosystem (Fig. 6.6; Rosenwax 2016). Rather than relying entirely on private developers, the university has undertaken the largest expansion seen in the university's 800 year history. The public institution made this possible by issuing bonds, which supplied the money to begin development. Further development will be funded selling parcels of land on long-term leases to developers under a carefully structured master plan. This will help fund infrastructure such as public transport and the delivery of water and power. A development body owned by the University manages the project. Urban planners in Australia should consider which stakeholders, other than universities can benefit from developments like this. To achieve wider benefits, government, industry and educators must be part of the project, recommends the Brookings Institution (Rosenwax 2016).

Fig. 6.6 University of Cambridge, United Kingdom. (Source: Rosenwax 2016)

6.11.2 Macquarie Park, Encouraging Collaboration

The area in northern Sydney is one of Australia's most significant innovation hotspots (Fig. 6.7). It is home to Macquarie University, a CSIRO centre, the 6000-person Optus campus and a concentration of major high technology and innovation businesses including Cochlear, Johnson & Johnson and Microsoft (Rosenwax 2016).

The innovation district will feature a range of initiatives designed to bring people together in "collision spaces". This includes creating start-up incubators, adding Wi-Fi to cafes to transform them into meeting spaces, encouraging walking and also providing a shuttle bus to make it easier for tenants to move around. However, key to success are the innovation programs such as hackathons, company engagement and solution providing events that mix start-ups, corporates, students and researchers.

6.11.3 South Lake Union, Seattle, Rezoning Catalyst

The decision to rezone land has been instrumental in the success of innovation ecosystems in the US and UK (Rosenwax 2016). Land rezoning can allow for choice space suitable for different types of businesses, along with housing to attract workers. The rezoning of under-utilised land, coupled with a clear vision for the location, provides certainty for investors. One example is South Lake Union in Seattle (Fig. 6.8), which was previously a light industrial area. It was only after the land was rezoned and attracted private investment that it became one of Seattle's fastest developing neighbourhoods. Amazon has established a new campus there and

Fig. 6.7 Macquarie Park, Sydney. (Source: Rosenwax 2016)

Fig. 6.8 South Lake Union. (Source: Rosenwax 2016)

Google will be moving in. The area also has a biotech and medical research community (Rosenwax 2016).

6.12 Conclusion

The new Western Sydney area, made up of existing centres and a new aerotropolis can take advantage of both bottom up and top down reform. Government investment in Western Sydney will drive investment confidence and this presents an opportunity for town centre revitalisation. It is the right time to be leveraging government-led initiatives such as the City Deals and galvanising communities through strong local leadership in the context of another 2 million people being added to Western Sydney by 2050.

It is also the right time for government to show leadership and support development in the region that is considered through sustainability and resiliency lenses. The landscape of the new Western City (The Parkland City) is located within the South Creek catchment. This is an opportunity to protect and enhance that corridor but to the patterns of the smaller tributaries to establish the framework for development.

References

City of Sydney (2017a) Community gardens. http://www.cityofsydney.nsw.gov.au/community/participation/community-gardens. Accessed 20 Apr 2017

City of Sydney (2017b) Sustainable Sydney 2030, community strategic plan 2017–2030. http://www.cityofsydney.nsw.gov.au/vision. Accessed 29 Apr 2017

Department of the Prime Minister and Cabinet (2017) Western Sydney city deal, smart cities plan, The Australian Government, Canberra. https://cities.dpmc.gov.au/22446/documents/44846. Accessed: 29 Apr 2017

Friends of the Highline (undated) The Highline. URL: http://www.thehighline.org/. Accessed 29 Apr 2017

Gittins R (2007) Microeconomic reform. http://www.rossgittins.com/2007/09/microeconomic-reform.html, Accessed 10 Apr 2017

Greater Sydney Commission (2016) Three cities. https://www.greater.sydney/three-cities. Accessed 5 Apr 2017

KaBOOM! (undated) Playful city USA. https://kaboom.org/playability/playful_city_usa. Accessed 20 Apr 2017

Katz B, Wagner J (2014) The rise of innovation districts, a new geography of innovation in America. https://www.brookings.edu/essay/rise-of-innovation-districts/. Accessed 5 Apr 2017

Lunsted P (2017) Tiny town of Girgarre in Victoria's north shows ingenuity in the face of job losses. http://www.abc.net.au/news/2017-04-29/girgarre-shows-ingenuity-in-the-face-of-job-loss/8462790, Accessed 29 Apr 2017

Rosenwax J (2016) Innovating, transforming Australian cities with innovation. http://www.aecom.com/content/wp-content/uploads/2016/09/AECOM_Transforming-Australian-cities-with-innovation.pdf. Accessed 5 Apr 2017

Wikipedia (undated) Microeconomic reform. https://en.wikipedia.org/wiki/Microeconomic_reform. Accessed 10 Apr 2017

Chapter 7
City Resilience and the Sydney Architect

Hugh Gardner and Georgia Vitale

> *One of the biggest challenges facing cities today is that no single organisation has complete control over the solutions to increase our resilience to the broad range of challenges we face.*
>
> Mayor George Ferguson, Bristol, UK

Abstract Unless cities acknowledge and focus on building resilience, shocks and stresses may cause decay or collapse with resulting local and global economic, social and environmental consequences. As part of planning and designing the third city in Western Sydney, we need to develop an understanding of how development scenarios enhance or undermine city resilience.

7.1 Introduction

Unless cities acknowledge and focus on building resilience, shocks and stresses may cause decay or collapse with resulting local and global economic, social and environmental consequences. As part of planning and designing the third city[1] in Western Sydney, we need to develop an understanding of how development scenarios enhance or undermine city resilience.

---

[1] The Greater Sydney Commission (GSC) contends that Greater Sydney must be reimagined as three cities. The three cities envisaged by the GSC are the established Eastern City, the developing Central City and emerging Western City. The third city planned for Sydney is known as the Western City containing the Western City Airport and the existing areas of Camden, Campbelltown, Liverpool and Penrith.

H. Gardner (✉) · G. Vitale
ARUP, Sydney, NSW, Australia
e-mail: hugh.gardner@arup.com; georgia.vitale@arup.com

R. Roggema (ed.), *Contemporary Urban Design Thinking*, Contemporary Urban Design Thinking, https://doi.org/10.1007/978-3-319-91950-8_7

The development of an international airport at Badgerys Creek within the next 10 years will come with population growth, transport infrastructure, new jobs and economic opportunities for the Western City. It is a unique chance to grow markets in international and domestic tourism, advanced logistics, aerospace industries freight, transport, health, education and the knowledge economy. The Western City will be home to over 1.5 million people and 400,000 jobs in 2036 (Greater Sydney Commission 2017).

Building resilience in the Western City will require an understanding of both what contributes to resilience and how it can be measured. The process of understanding what city resilience means for Sydney has started and has momentum, so now is the time to consider how to act.

> The West District is Greater Sydney's bushland city, a place surrounded by World Heritage-listed landscapes, with a sprinkling of towns and centres that combine village charm and heritage character. This iconic landscape is more than a backdrop – it is the West District's underlying asset. People travel from around the world to experience this landscape – mountains, escarpments, rural hinterland and rivers. The landscape is the foundation for how we think and plan for the District – it resonates in our places and informs the District's design and structure". Lucy Hughes Turnbull AO Chief Commissioner Greater Sydney Commission (GSC 2017)

There's an emerging understanding of what city resilience means in Sydney.

Sydney has participated in the Rockefeller Foundation's 100 Resilient Cities program and there is now a common definition of city resilience and a framework to guide decision-making. Ministerial priorities have placed resilience high on Greater Sydney's agenda.

In the (former) Minister's speech (NSW Government, Parliament of NSW 2015), introducing the Greater Sydney Commission Bill 2015, six objectives of the bill are highlighted. The 5th objective is resilience planning, aimed at encouraging development that is resilient and mindful of natural hazards to ensure that the city is prepared to address issues resulting from hazards and changes of climate. The Commission will develop district plans that will incorporate city resilience responses.

The Western City Draft District Plan (2017) notes that the necessary actions to build such resilience include:

- Appropriately equipping and educating communities;
- Completing a Floodplain Management Review of the Hawkesbury-Nepean Valley;
- Undertaking 'hazard-mapping', considering findings throughout land use planning processes.

The reality is that Sydney is subject to a broader set of natural and man-made pressures that have the potential to cause significant disruption, at worst leading to social breakdown, economic decline or physical collapse. Architects and urban designers of the future will be part of the response.

Resilience planning takes a city systems perspective, enabling a better understanding across infrastructure, institutions, knowledge and ecosystems, and importantly across city boundaries. If government policy makers, NGOs and the private sector are to contribute to a more resilient Sydney – in particular at the third city – they need to explore and understand the factors that contribute to resilience at a city scale. Importantly, this requires a detailed discussion about the networks of control and influence that reach beyond administrative boundaries.

It is the right time for Sydney to have a serious discussion about city resilience because no single action will make Sydney resilient to shocks and stresses. Resilience will be built through collective actions, across spatial boundaries, and across levels of government with the help of built environment professionals.

7.2 Resilience Thinking

It is useful to first explore where the concept of city resilience has come from and why it finds application across disciplines or faculties. Cities can be viewed as systems comprising various elements. When combined these systems have qualities that may not be present individually. Changes are systemic and dynamic. A system's behaviour can only be understood by looking at the entire system and not its elements in isolation.

7.2.1 From Ecology to City Planning and Design

Emerging from the field of ecology in the 1970s, 'resilience' was understood as the capacity of a system or agent to maintain or recover functionality in the event of disruption or disturbance. Work by Holling (1973) established one entry point as the ecological systems perspective (Carpenter et al. 2001), which describes resilience as:

- The ability of a system to absorb disturbance and maintain function;
- The capacity of the system to self-organise (versus lack of organization, or organization forced by external factors);
- The degree to which the system can build and increase the capacity for learning and adaptation.
- As a result, the concept of resilience is commonly applied to systems rather than individual units.

Early this century it was recognised that resilience should include the overarching goal of a system to continue to function to the fullest possible extent in the face of stress to achieve its purpose, where resilience is a function of both the vulnerability of the system and its adaptive capacity (Dawley et al. 2010; Da Silva et al. 2012).

The systems approach recognised physical and non-physical components of cities. It is important to understand scale as a framework for city systems and the interdependencies between them. City systems operate at multiple scales and these

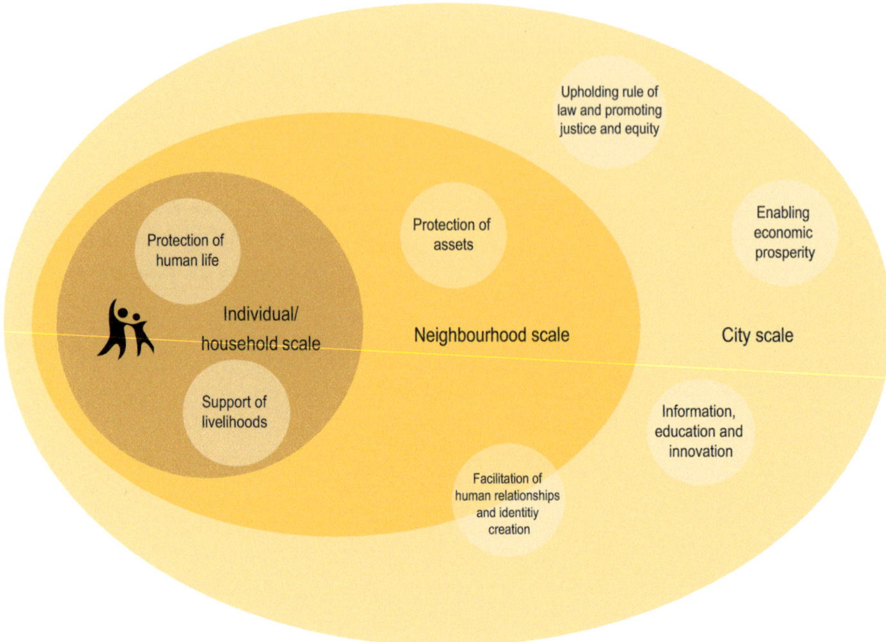

Fig. 7.1 City resilience and scale. (Produced by Arup)

scales do not fit into a linear hierarchy, but we can represent them in simplified forms similar to those in Fig. 7.1.

The challenge is that resilience at a city scale cannot be measured directly, other than in terms of changing performance of systems in response to shocks and stresses. Networks, such as infrastructure, institutions, ecosystems and knowledge, that make up city systems will, in a resilient city, demonstrate a number of key characteristics that can be used to measure progress.

City resilience was in its early conception focussed on the challenge of change in climate (Da Silva et al. 2012; Asian Cities Climate Change Resilience Network 2013; Kernaghan and Da Silva 2014; Asian Development Bank 2014). It was argued that the traditional risk assessments founded on spatial analysis and climate projections failed to recognise the uncertainty of climate change data or the complexityof cities (Da Silva et al. 2012).

The resulting alternate approach looked at the problem from an urban systems perspective, founded on an understanding of how the city functions, its boundaries of control and influence and how it could continue to function in the face of a wide range of shocks and stresses (Da Silva et al. 2012).

In a domestic iteration of the concept, the Trusted Information Sharing Network (TISN) for Critical Infrastructure Resilience was established as Australia's primary national engagement mechanism for business-government sharing and resilience building initiatives on critical infrastructure resilience. TISN developed a Critical Infrastructure Resilience Strategy in order to ensure the continued operation of critical infrastructure in the face of all hazards (Australian Government 2010). Resilient critical infrastructure aims to ensure continued provision of essential services to businesses, governments and communities.

Resilience has made its way into discussions about governance. Of particular interest is city governance that refers to the range of institutions and agencies involved in governing (Lebel et al. 2006; Walker and Cooper 2011; Leichenko 2011).

Economic resilience is yet another established area of research (Rose 2004; Simmie and Martin 2010; Hudson 2010; Davies 2011; Martin 2012; Leichenko and Thomas 2012). At its core the concept is currently dealing with the ability of city economies to withstand external pressure, to respond positively to external change, and demonstrate longer-term adaptability or learning capacities (Hudson 2010; Davies 2011).

All these concepts of resilience identify a need to anticipate external events and to respond flexibly and efficiently. Lebel et al. (2006) concludes that the best responses will come from places with decentralised and accountable institutions that operate in a transparent and inclusive manner.

Public participation has entered the city resilience narrative because of its role in building trust and shared understanding. The idea is that local knowledge and local monitoring facilitate interaction and feedback, while accountable and just institutions pursue socially just distributions of benefits and risk among the populations they serve (Lebel et al. 2006).

7.2.2 *What We Mean by City Resilience*

What is common about the different perspectives on resilience is the focus on how you should approach the challenge of ensuring assets (social, ecological, physical, economic) are available for communities, rather than what specifically you should do. That is communities should aim to:

- Function/ maintain core functions under stress (Holling 1973; Carpenter et al. 2001; NEMC 1999; Maguire and Cartwright 2008; Rose 2004);
- Build strong/robust critical infrastructure (O'Rourke 2007; Arup 2011);
- Learn and re-organise from change (Carpenter et al. 2001; Arup 2011; Folke 2006);
- Transform to a more resilient state as a result (Plodinec 2009; Maguire and Cartwright 2008).

The Rockefeller Foundation and Arup contention is that resilience in the cities context does not refer to 'bouncing back' to a pre-existing state (Holling 1973), but rather the ability of institutions, infrastructure, ecosystems and knowledge networks

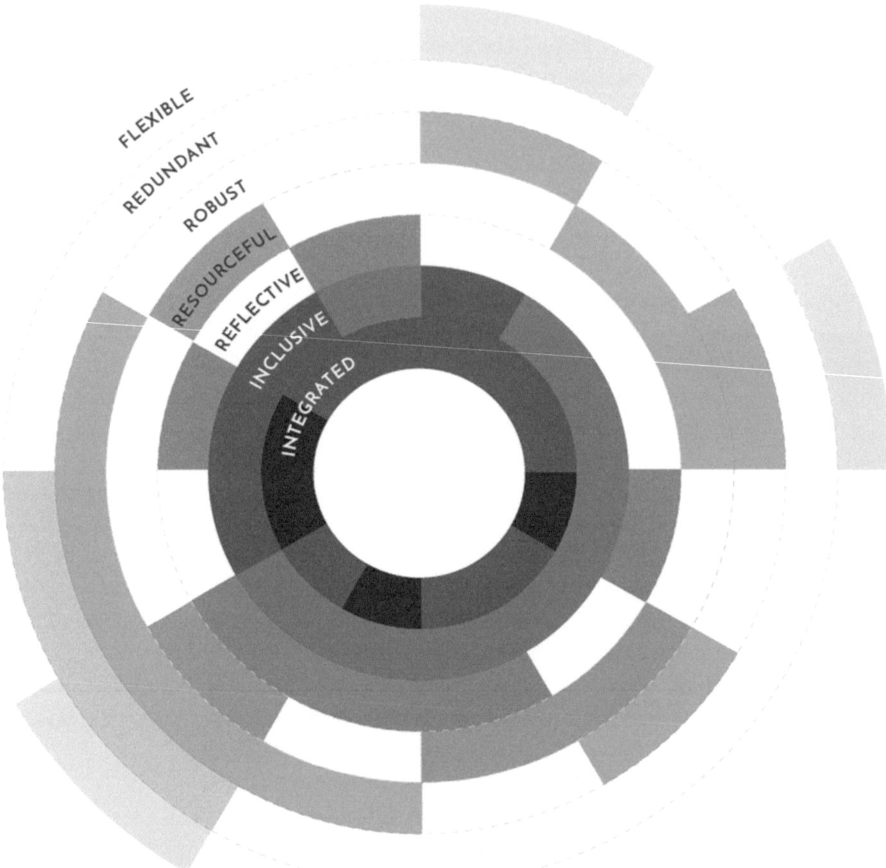

Fig. 7.2 Characteristics of resilient cities. (Produced by Arup)

to evolve and adapt so that urban populations can survive and thrive when faced with unpredictable shocks and stresses (Arup and The Rockefeller Foundation 2014).

The definition of city resilience is: "the capacity of individuals, communities, institutions, businesses, and systems within a city to survive, adapt, and grow no matter what kinds of chronic stresses and acute shocks they experience" (Arup and The Rockefeller Foundation 2015). Resilient cities have seven distinct characteristics. They are:

- Inclusive: Inclusion emphasises the need for broad consultation and engagement of communities, including the most vulnerable groups.
- Integrated: Integration and alignment between city systems promotes consistency in decision-making and ensures that all investments are mutually supportive to a common outcome.

- Reflective: Reflective systems are accepting of the inherent and ever-increasing uncertainty and change in today's world.
- Robust: Robust systems include well-conceived, constructed and managed physical assets, so that they can withstand the impacts of hazard events without significant damage or loss of function.
- Flexible: Flexibility implies that systems can change, evolve and adapt in response to changing circumstances.
- Redundant: Redundancy refers to spare capacity purposely created within systems so that they can accommodate disruption, extreme pressures or surges in demand.
- Resourceful: Resourcefulness implies that people and institutions are able to rapidly find different ways to achieve their goals or meet their needs during a shock or when under stress (Fig. 7.2).

Further detail can be found in the City Resilience Index (Arup and The Rockefeller Foundation 2015). The City Resilience Index is based on 3 years of research contributions, case studies and pilot schemes conducted in diverse cities across the world. It is a design framework that will be useful during the formative stage of the Western City in Sydney. It encourages city planners and designers to think of the built environment by exploring:

- People: The health and well-being of everyone living and working in the city;
- Organisation: The systems within the economy and society that enable urban populations to live peacefully, and act collectively;
- Place: The quality of infrastructure and ecosystems that protect, provide and connect us;
- Knowledge: Appropriate leadership and strategy enabling the city to learn from the past and take timely action.

This work emphasises resilience as a concept for dealing with change and learning in urban environments, rather than as system breakdown and return to stability.

Its purpose is to inform urban planning and design practice and investment patterns, to better enable urban communities – particularly disadvantaged communities – to survive and thrive following significant social, environmental, or economic stress and disruption.

The City Resilience Index (2015) is currently the most widely used assessment for city resilience.

7.3 Sydney's Resilience Journey

Sydney's resilience journey has evolved from an Integrated Regional Vulnerability Assessment model considering climate impacts such as flooding, coastal erosion and inundation and bush fires and heat waves (Jacobs et al. 2014a, b) to the 100 Resilient Cities initiative, known as Resilient Sydney (100 Resilient Cities 2016a).

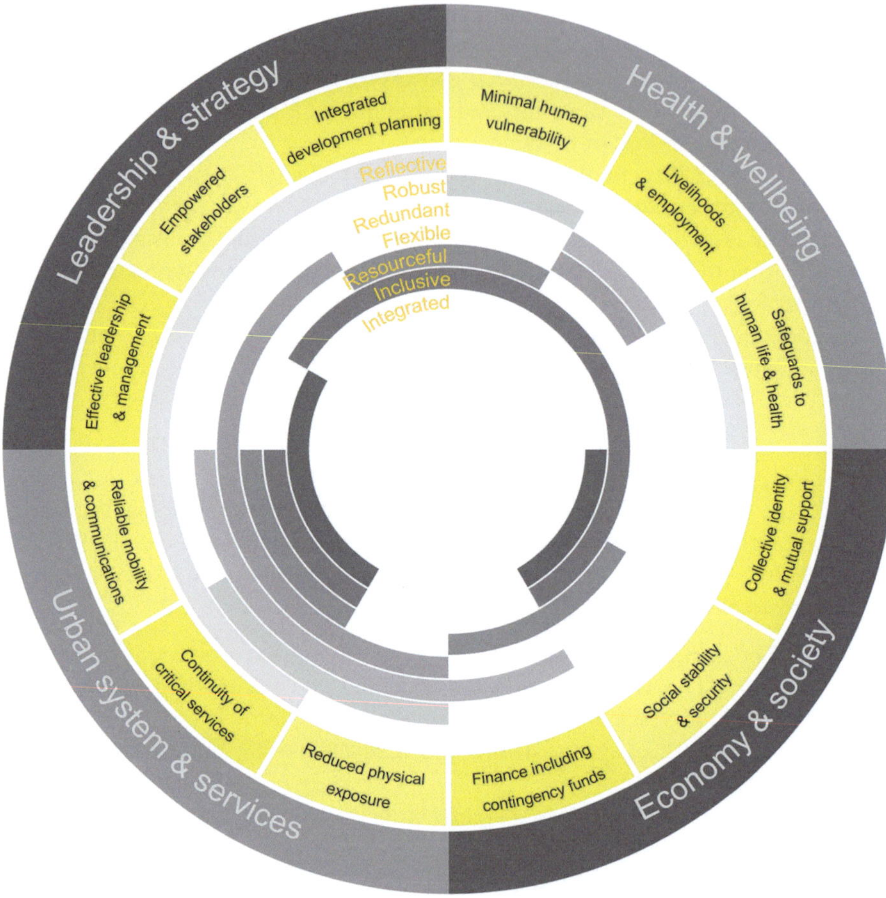

Fig. 7.3 The city resilience index

Pioneered by the Rockefeller Foundation, Sydney as part of the 100 RC network is working in collaboration with the City of Sydney, the metropolitan councils of Sydney and the NSW state government. Resilient Sydney uses the City Resilience Framework (Arup and The Rockefeller Foundation 2015), summarised in Fig. 7.3, to enable city governments to assess the current status of their city system and measure progress against four essential dimensions of urban resilience:

- Health and wellbeing – everyone living and working in the city has access to what they need to survive and thrive;
- Economy and society – the social and financial systems that enable urban populations to live peacefully and act collectively;
- Leadership and strategy – the processes that promote effective leadership, inclusive decision-making, empowered stakeholders and integrated planning;

- Infrastructure and environment – the man-made and natural systems that provide critical services, protect, and connect urban assets enabling the flow of goods, services and knowledge.

Since being selected to join the 100 Resilient Cities network, Sydney's Chief Resilience Officer has worked with businesses, government, academia, the social services, non-profit sector and the community to identify the key resilience challenges facing Sydney. The assessment includes identification of the top shocks and stresses for Sydney now and into the future (100 Resilient Cities 2016b):

Shocks	Stresses
Extreme weather (heat, bushfire, storms, flood)	Increasing health service demand
Financial institution failure	Diminishing social cohesion
Infrastructure failure	Loss of housing affordability
Disease pandemic	Increasing chronic illnesses
Water crisis	Lack of transport diversity
Digital network failure	Insufficient employment density
Cyber attack	Increasing geographic inequity
Terror attack	Rise in drug and alcohol abuse

Resilient Sydney has progressed Sydney's resilience journey from a natural hazards risk based approach focusing purely on climate change to a broader view of how Sydney can manage disruptions to the function of the city and to create opportunities in adapting to future challenges. It will provide the strategy for implementing Sydney's first fully integrated response to city resilience.

7.4 Resilience in the Third City

High quality strategic planning and design is the number one resilience solution identified by cities in the 100 Resilient Cities program (100 Resilient Cities 2016b). Sydney's resilience challenges need to be understood within the context of Western Sydney given the localised nature of shocks and stresses. Every community faces different threats and conditions and there is no one-size-fits-all approach. To localise challenges and opportunities it is important to consider (Carpenter et al. 2001; Lebel et al. 2006; Vale 2014):

- Resilience to what?
- Resilience of what part of the city system?
- Resilience for whom?

Equally challenging in the context of the Western City is the degree of uncertainty in planning for a long-term horizon whereby a number of alternative models of growth over time may play out and consequently impact on those three questions.

7.4.1 Resilience to What?

As the Western City emerges, Sydney's long term stresses are an important consideration. The Western City not only needs to understand and address the potential for impact of these stresses, but also needs to be part of the response for Greater Sydney. Settlement patterns and design in the Western City will clearly have an impact beyond the designated geographic area and residential community – it could help build resilience across the city region.

From a climate change perspective, the Western City already experiences more hot days than other part of Greater Sydney and exposure to extreme heat will be an ongoing challenge, as will risks of flooding, increases in the frequency and intensity of storm events and rainfall. Planning, and design in the third city needs to consider the long-term impacts of climate change on infrastructure performance and community resilience.

With a business as usual approach, resilience stresses for the Western City could include insufficient employment density and an inability to realise agglomeration benefits, which would lead to an increase in geographic inequality across Greater Sydney. The planning and design response needs to recognise that the employment mix will change in the Western City over the short to medium term and there will be a need to transition to a more knowledge intensive skills base.

A lack of transport diversity and an overreliance on road based transport would particularly impact vulnerable communities who are more susceptible to fluctuations in factors such as petrol prices and interest rates. Increasing chronic illness and demand for health services are amongst the negative impacts connected to car based planning and design (Newman and Kenworthy 1989).

7.4.2 Resilience of What Part of the City System?

Preservation and management of city systems are fundamental to city resilience and wellbeing: basic needs for survival, security, health, good social relations and esteem and freedom of choice and action (Maslow 1971; Alcamo et al. 2003; Huitt 2004).

In the Western City, it will be useful to establish a simple model that leads to focus areas. A useful model is developed in the article. A systems approach to meeting the challenges of urban climate change (Da Silva et al. 2012) and it identifies three types of network within the city system: infrastructure networks, knowledge networks and institutional networks. The urban condition includes additionally the outcome of cities – wellbeing – and the resources that enable it – ecosystems (Fig. 7.4).

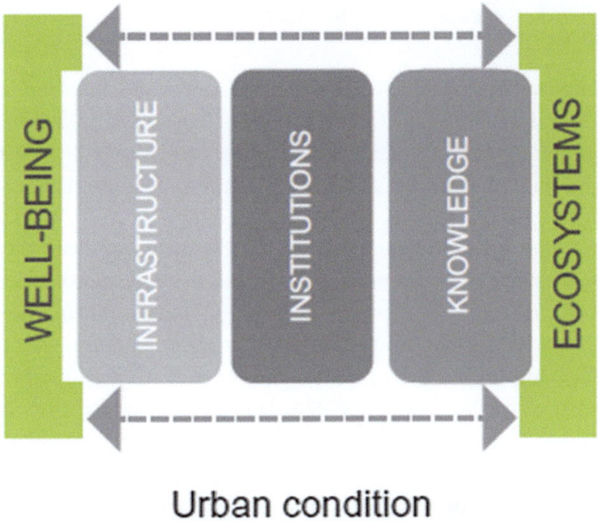

Fig. 7.4 Simplified urban system model. (Da Silva et al. 2012)

A brief description of the model components:

- Infrastructure networks: the key physical and technological asset of the city,
- Knowledge networks: the structures that allow information to be created and shared
- Institutional networks: the governance, economic and social frameworks that influence human interaction
- Well-being: basic human needs (food, water and shelter), access to goods and livelihood opportunity, security, health, social relations and freedom to act (Maslow 1971; Alcamo et al. 2003; Huitt 2004)
- Ecosystem services: benefits to people including provisioning, regulating, supporting and cultural.

We know resilience will only be built by multiple and cumulative contributions over time. Within the model, the Western City could decide to focus on interventions in one or all of the three networks in the simplified urban system model. The current NSW Government approach (Greater Sydney Commission2017) appears to lean towards establishing infrastructure based (primarily movement) support, however, the formal education networks (primarily schools and universities) and business agglomeration potential around movement infrastructure have been other focuses of strategic planning.

The Western Sydney City Deals framework (Australian Government 2016) provides an opportunity for more integrated city governance and new funding opportunities with new institutional dynamics. The focus areas are providing or attracting investment to develop infrastructures, increased employment diversity, affordable housing solutions, environmental/liveability outcomes and a reduction in uncertainty associated with decision making.

7.4.3 *Resilience for Whom?*

Given the growth projections and proposed land use change in the Western City Draft District Plan (Greater Sydney Commission 2017), the population profile – the people, community forces and businesses or working opportunities – will change dramatically. Understanding this future community profile and their ability to contribute to the success of the city, manage disruption and take advantage of opportunities is critical. Equally important is thinking more creatively about the trends that will shape the Western City community and to consider scenarios beyond business as usual extrapolations of population and demographic projections.

An understanding of the city context and the systems that support the Western City highlights the potential shocks and stresses likely to be faced by communities west of the Western City. The long term stresses of land use changes associated with productive lands and Sydney's food bowl are trans-boundary issues and emphasise that the question of resilience for whom extends beyond the planning boundary and is of consequence to the city as a whole. Equally the Western City could lead the way in building city resilience for Greater Sydney.

Resilience thinking provides us with an established framework (summarised in Fig. 8.3) to consider these questions/challenges and where and how we want to intervene.

7.5 The Role of the Urban Designer in City Resilience

Planning and design approaches have always been altered by natural hazard events. The built environment industry has most recently been trying to implement a hazard and risk based approach through lessons learned, particularly with a view to addressing climate risks. For example: at a very practical level designers (mostly in Western Europe) are responding in areas subject to storm surge and sea-level rise by rethinking landscaping. Interventions include berms, wetlands, mangroves and swales with native plants as natural barriers that reduce wave energy and absorb water (Brandes and Alice 2013).

However, with a greater understanding of the city as a system, the role of the urban designer in contributing to city resilience is fundamental. Strategic urban planning has been identified as the number one resilience solution by cities in the 100 Resilient Cities program. City planners and urban designers are well placed to influence Sydney's resilience as seen by "the new school of city planning and design (which) embraces complexity and uncertainty and this is precisely where resilience has found its place" (Williams and Kernaghan 2015).

Fundamentally, the designer's role in helping Sydney bounce forward requires them to:

- Contribute more openly and meaningfully to civic dialogue
- Apply a resilience thinking lens as part of any design response (in particular at the city scale)

- Identify interdependent systems given resilience plays out at multiple scales
- Identify the critical gaps, where action and investment to build resilience will be most effective, or where deeper analysis or understanding is required
- Think beyond architecture-as-spectacle: focus on an integrated and inclusive process
- Design for flexibility by accommodating a full range of alternative futures, while giving focus to the fundamental interventions required and how they exist in time.

Design plays a critical role in achieving the aspiration we have for our future because good design is about deep understanding and a creative synthesis of ideas, issues and people. Design offers a stand-alone and contributing process to planning our future. Importantly, design is an iterative and inclusive process with much to offer to decision making and planning in government.

The Government Architect NSW has developed Better Placed, an architecture and design policy for NSW (Government Architect NSW 2017). The policy establishes objectives and expectations of design in creating good places, provides principles and direction to achieve these and provides a framework for examining places and reviewing proposals, from a design perspective. The policy articulates well the role of designers in the process of achieving liveable, productive and resilient environments. It is an emerging discussion and one that needs contribution from students, academics and practitioners across the built environment disciplines.

> Understanding resilience in metropolitan Sydney is important work for planning our future city. I am delighted to see progress through the 100 Resilient Cities program and look forward to working with the Greater Sydney Commissionand councils to improve the resilience of Sydney. Rob Stokes MP, (former) NSW Minister for Planning. (100 Resilient Cities 2016b)

References

100 Resilient Cities (2016a) Preliminary resilience assessment. Retrieved from http://www.cityofsydney.nsw.gov.au/__data/assets/pdf_file/0005/263975/2016-503932-Report-Resilient-Sydney-PRA-FINAL-ISSUED.pdf

100 Resilient Cities (2016b) City context report. Retrieved from http://www.cityofsydney.nsw.gov.au/__data/assets/pdf_file/0006/263976/2016-503932-02-Report-Resilient-Sydney-City-Context-Report-FINAL-ISSUED.pdf

Alcamo J, Hassan R, Scholes R, Ash N, Carpenter S, Pingali P, Bennett E, Zurek M, Chopra K, Leemans R (2003) Ecosystems and human well-being: a framework for assessment (Millennium ecosystem assessment). Island Press, Washington, DC

Arup (2011) International Federation of the Red Cross and Red Crescent Societies Community Based Disaster Risk Reduction Study – characteristics of a safe and resilient community

Arup and The Rockefeller Foundation (2014) City resilience framework. Retrieved from https://assets.rockefellerfoundation.org/app/uploads/20140410162455/City-Resilience-Framework-2015.pdf

Arup and The Rockefeller Foundation (2015) City resilience index. Retrieved from https://www.arup.com/publications/research/section/city-resilience-index

Asian Cities Climate Change Resilience Network (2013) ACCCRN city projects. The Rockefeller Foundation Asia Office
Asian Development Bank (2014) Urban climate change resilience: a synopsis. Retrieved from https://www.adb.org/sites/default/files/publication/149164/urban-climate-change-resilience-synopsis.pdf
Australian Government (2010) Critical infrastructure resilience strategy. Retrieved from https://www.tisn.gov.au/Documents/Australian+Government+s+Critical+Infrastructure+Resilience+Strategy.pdf
Australian Government (2016) Western Sydney city deal factsheet. Retrieved from https://cities.dpmc.gov.au/western-sydney-city-deal
Brandes U, Alice L (2013) Risk & resilience in coastal regions. Urban Land Institute, Washington, DC
Carpenter S, Walker B, Anderies J, Abel N (2001) From metaphor to measurement: resilience of what to what? Ecosystems 4:765–781
Da Silva J, Kernaghan S, Luque A (2012) A systems approach to meeting the challenges of urban climate change. Int J Urban Sustain Dev 4(2):1–21
Davies S (2011) Regional resilience in the 2008–2010 downturn: comparative evidence form European countries. Camb J Reg Econ Soc 4:362–389
Dawley S, Pike A, Tomaney J (2010) Towards the resilient region? Local Econ 25:650–667
Folke C (2006) Resilience: the emergence of a perspective for social-ecological systems analyses. Glob Environ Chang 16:253–267
Government Architect NSW (2017) Better placed: an integrated design policy for the built environment of New South Wales. NSW Government
Greater Sydney Commission (2017) Draft west district plan. Retrieved from https://www.greater.sydney/western-city-district
Holling C (1973) Resilience and stability in ecological systems. Annu Rev Ecol Syst 4:1–24
Hudson R (2010) Resilient regions in an uncertain world: wishful thinking or a practical reality? Camb J RegEcon Soc 3:11–25
Huitt W (2004) Maslow's hierarchy of needs. Educational psychology interactive. Available from: http://ways-ahead.net/meditation/037-2-Maslow.pdf
Jacobs B, Boronyak L, Dunford S, Kuruppu N, Lewis B, Lee C (2014a) Towards a resilient Sydney – supporting collective action to adapt sub national government services to regional climate change. Proceedings of the third international conference on climate change and social issues
Jacobs B, Lee C, O'Toole D, Vines K (2014b) Integrated regional vulnerability assessment of government services to climate change. Int J Clim Chang Strateg Manag 6(3):272–295
Kernaghan S, da Silva J (2014) Initiating and sustaining action: experiences building resilience to climate change in Asian cities. Urban Clim 7:47–63
Lebel L, Anderies J, Campbell B, Folke C, Hatfield-Dodds S, Hughes T, Wilson J (2006) Governance and the capacity to manage resilience in regional social-ecological systems. Ecol Soc 11:19
Leichenko R (2011) Climate change and urban resilience. Curr Opin Environ Sustain 3:164–168
Leichenko R, Thomas A (2012) Coastal cities and regions in a changing climate: economic impacts, risks and vulnerabilities. Geogr Compass 6:327–339
Maguire B, Cartwright S (2008) Assessing a community's capacity to manage change: a resilience approach to social assessment. Australian Government Bureau of Rural Sciences, Canberra
Martin R (2012) Regional economic resilience, hysteresis and recessionary shocks. J Econ Geogr 12:1–32
Maslow A (1971) The farther reaches of human nature. Viking Press, New York
NEMC (1999) Using a performance-based measurement system. Proceedings 14th Annual WTQA symposium. Arlington: 13–15 July 1998
Newman P, Kenworthy J (1989) Cities and automobile dependence: an international sourcebook. Gower, Aldershot

NSW Government, Parliament of NSW (2015) Greater Sydney commission bill 2015. Retrieved from https://www.parliament.nsw.gov.au/bills/Pages/bill-details.aspx?pk=919

O'Rourke TD (2007) Critical infrastructure, interdependencies, and resilience. The Bridge 37(1):22–29

Plodinec M (2009) Definitions of resilience: an analysis. Community and Regional Resilience Institute (CARRI) Retrieved from http://resilientus.org/library/CARRI_Definitions_Dec_2009_1262802355.pdf

Rose A (2004) Defining and measuring economic resilience. Disaster Prev Manag 13(4):307–314

Simmie J, Martin R (2010) The economic resilience of regions: towards an evolutionary approach. Camb J Reg Econ Soc 3:27–44

Vale LJ (2014) The politics of resilient cities: whose resilience and whose city? Build Res Inf 42(2):191–201

Walker J, Cooper M (2011) Genealogies of resilience: from systems ecology to the political economy of crisis adaptation. Secur Dialogue 43:143–160

Williams T Kernaghan S (2015) Does metropolitan governance bring us closer to resilient cities? The fifth estate. Retrieved from https://www.thefifthestate.com.au/columns/spinifex/does-metropolitan-governance-bring-us-closer-to-resilient-cities

Chapter 8
A Green Grid for Sydney's Growth

Barbara Schaffer

Abstract Sydney's population is forecast to increase 80% by 2054. An additional three million people will live and work in metropolitan Sydney. As population density increases, our challenge is to shape the built environment so as to ensure that Sydney remains one of the worlds most distinctive and liveable cities (Fig. 8.1).

8.1 Introduction

Sydney's population is forecast to increase 80% by 2054. An additional three million people will live and work in metropolitan Sydney. As population density increases, our challenge is to shape the built environment so as to ensure that Sydney remains one of the worlds most distinctive and liveable cities (Fig. 8.1).

In acknowledging that green space is a key hallmark of liveability, the Government Architect NSW (GANSW) proposed the creation and consolidation of a network of high-quality green areas that connect town centres, public transport networks and major residential areas. Now known as the Sydney Green Grid, and regarded as an integral part of the Greater Sydney Region Plan (Greater Sydney Commission 2017a) and the most recent District Plans for Sydney (Greater Sydney Commission 2017b). This network aims to anchor sustainable development while maximising quality of life and wellbeing (Fig. 8.2).

Although Sydney has many green and water enriched spaces, an overarching schema that approaches them in a connected way is missing – and this is the aim of this project – ensuring that their contribution to our quality of life, the environment and the economy are maximised, rendering a working-whole that is far greater than the sum of its parts (Fig. 8.3).

Some places are already well served, and here the task is making sure it remains of high quality and is managed effectively, while other places suffer from a lack of

B. Schaffer (✉)
Government Archtect NSW, Sydney, NSW, Australia
e-mail: barbara.schaffer@planning.nsw.gov.au

R. Roggema (ed.), *Contemporary Urban Design Thinking*, Contemporary Urban Design Thinking, https://doi.org/10.1007/978-3-319-91950-8_8

Fig. 8.1 Map of Metropolitan Sydney

green space and here we need to extend the current network and create new open space where it is lacking.

8.2 Green Infrastructure

The Sydney Green Grid is a green infrastructure, design-led strategy that includes the full range of open spaces: from national, regional and local parks through the harbour, ocean beaches, wetlands, rivers and creeks, to playgrounds, playing fields, golf courses and cemeteries. Furthermore, interconnected linkages are fostered within the wider public realm through enhancing creek corridors, transport routes, suburban streets, footpaths and cycle-ways. The Green Grid is therefore an open-space interconnecting network that will keep the city cool, encourage healthy life-styles, enhance biodiversity and ensure ecological resilience (Fig. 8.4).

Green infrastructure is an asset, as integral to Sydney's metabolism as its roads, rail lines and storm water pipes, and valued for the whole range of social, health, environmental, economic and educational benefits it brings to Sydney. The importance of an integrated approach to management, enhancement and extension is paramount. Green space is an infrastructure asset that requires the same kind of protection, investment and innovation we afford more familiar types of built infrastructure.

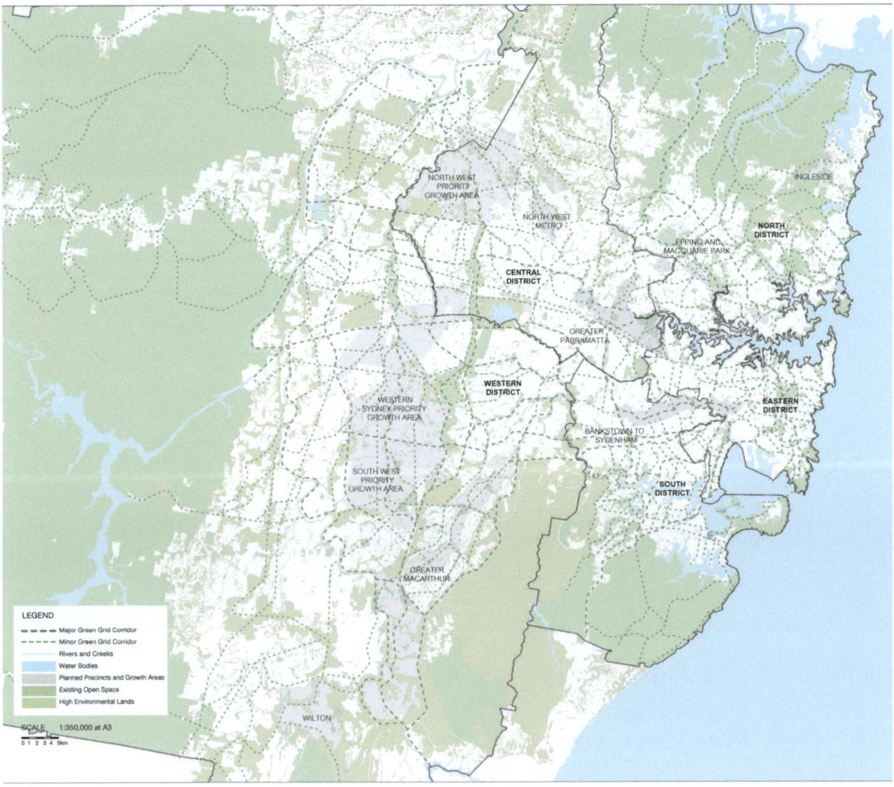

Fig. 8.2 Map of the Metropolitan Green Grid Network. (Source: GANSW and Tyrrellstudios 2016a)

Green infrastructure has an increasingly important role to play as we face the challenges of population and urban growth on the one hand and climate change on the other. Well-designed and planned green infrastructure will help absorb flood water, cool the urban environment, clean the air, provide space for local food production and ensure the survival of Sydney's fauna and flora as well as providing space for recreation, sport and leisure.

The Sydney Green Grid underscores the value of green and open space as pivotal to the choices we make when promoting economic growth, health and wellbeing. As a network, it will provide links and connections between places, encourage walking and cycling, highlight landscape and heritage, and support local economies. By providing informal places for people to visit and interact, social capital is both created and enhanced. Future investment in parks and recreation will play a vital role in Sydney's ability to attract business and create jobs.

Fig. 8.3 Aerial view of Sydney Harbour to Botany Bay

8.3 How Was the Green Grid Initiated?

Inspired by the All London Green Grid (Greater London Authority 2012), GANSW undertook a pilot project in Parramatta where the provision of open space at regional, district and local levels was assessed. Open space deficiencies were interrogated and determined where additional open space was required to ensure equity of access. In doing so an interconnected network of open space was proposed for Parramatta. The creek corridors were looked at and it was illustrated how this network could be used for flood management as well as for walking and cycling. Furthermore, the street networks and potential infrastructure corridors such as light rail were examined, and it was suggested how the unique heritage components of the city could be integrated into the living experience of Parramatta.

Having demonstrated the value of applying this thinking at the district and metropolitan scales, the GANSW has worked close with the Greater Sydney Commission to create an evidence-based, open-space audit across Metropolitan Sydney as a baseline for exploring opportunities to create an interconnected Metropolitan network that will support the projected population, housing and employment targets.

The Sydney Green Grid posits an economic case for investment in green infrastructure beyond the provision of open space for recreation alone. Consequently, it is now understood by many arms of urban planning and design in both the public

Fig. 8.4 Diagram Green Infrastructure. (Source: GANSW 2017)

and private realm as a value proposition, where benefit outweighs cost by raising the overall quality of investment in both the open space and infrastructure alike.

As similar initiatives are happening around the globe, the ideas were benchmarked against thinking from around the world, where the benefits for example from reduced flood risk, improved health outcomes and increased ecological resilience are being quantified and where the concept of green infrastructure is becoming mainstream. For example, in Chicago greening only a small percentage of the city's rooftops has significantly reduced air pollution. Chicago estimates that this investment could result in avoided health costs of 29 to 111 million dollars annually (Chicago City Plan Commission 2008).

In 2011 Philadelphia created the Green City Clean Waters program, a 25 year, 2.5-billion-dollar plan to protect and enhance the city's catchments by managing storm water with innovative green infrastructure. The city estimates that using green infrastructure in lieu of traditional approaches could save 8 billion dollars over the life of the program (Philadelphia Water Department 2011).

Based on Sydney's unique character, the aim is to articulate a clear and compelling economic rationale for future investment in green infrastructure. The research uncovered a number of cost/benefit analyses, but although well intended, they are limited in scope as the implementation is fragmented.

Sydney's current parks and open spaces and other green features, such as street trees are managed by over 40 different public authorities and agencies (GANSW and Tyrrellstudios 2016a). This results in an approach to planning and management of a suite of assets that does not fully realise the potential that could be harnessed by a more integrated approach. To successfully create green infrastructure at the landscape scale, conventional 'silo' modes of thinking within government must be transcended, promoting an approach to planning, design and implementation where key agencies work together to support a common goal. This strategy needs to be owned by many. As a leading example, the Greater Sydney Commission is actively promoting and advocating the Green Grid across government, influencing planning strategies, local government open space network plans, and urban transformation precincts.

The delivery of the Sydney Green Grid vision is a complex and challenging task. It is a long-term, evolutionary process that will require bipartisan support on all levels of government – national, regional and local. It is important that the strategic Green Grid projects identified in the District Plans can be delivered through the planning and development process, beginning with appropriate policies by local government in their Local Environment Plans.

The Green Grid recognises that there are significant benefits to be gained from the investment in a network of interlinked and high-quality public open spaces. Connecting previously isolated open spaces encompassing town centres, major employment hubs, public transport nodes and major residential areas, brings together the area's urban and natural elements.

The Green Grid provides the context for development, builds community capacity for change through positive environmental improvements and enhances land values. The longer-term vision for the Green Grid is to improve the quality, quantity and performance of open space to meet the social, recreational, health and transport needs of a significantly increased resident and working population (GANSW and Tyrrellstudios 2016a).

By combining multiple layers of spatial information, the Green Grid has created a cohesive map of green assets across metropolitan Sydney (ibid.). It provides the opportunity to overlay population, infrastructure, existing vegetation and cycleways with biodiversity features in the Sydney Basin.

The data sets include hydrological, ecological, agricultural and recreational grids. These grids are fragments of information when viewed separately, but when integrated, they enable a holistic reading of Sydney's green and open space. In this sense, the Green Grid has become a collective tool that is useful for multiple agencies (Fig. 8.5).

The project reinforces the primacy of open space in metropolitan planning. The Green Grid has informed policy and planning strategies, and has provided the spatial framework for the development of physical projects. Rather than green assets being an afterthought, the Green Grid is driving, shaping and framing Sydney's green space. Implementation of the green grid is ongoing, with strategies being executed at a number of different scales of the city including district, city and local scales.

8.4 Green Grid Projects by Scale

The Green Grid works at multiple scales all at once: from regional and district planning to Local Government Area (LGA) initiatives, to projects for local sites, the Green Grid strategies can be applied to create positive interventions. The multi-layered sophistication that the Green Grid framework offers has enabled collaborative engagement in planning strategies.

The Green Grid has helped inform the Greater Sydney Commission's focus of Sydney as three cities (Fig. 8.6), with unique characteristics and needs – an approach where place is framed by location as well as landscape. The Greater Sydney Commission identifies (Greater Sydney Commission 2017a):

- The eastern city (the Harbour City);
- The central city (the River City);
- The western city (the Garden City).

8.4.1 District: Central City District

At the district level, several documents reveal the potential of the Green Grid. For the Central City district, the Open Space Audit Report and the Central City Spatial Framework were prepared to inform the preparation of the District Plan.

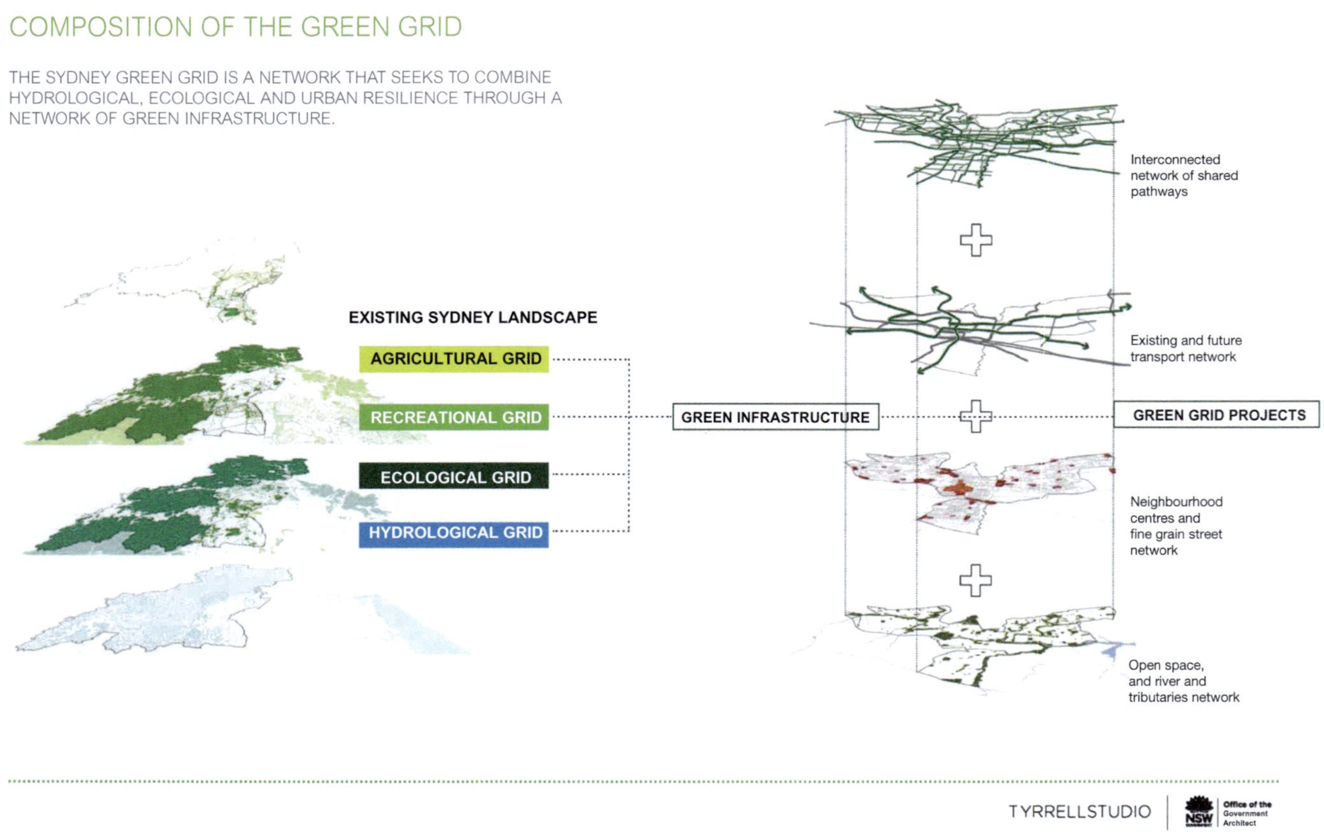

Fig. 8.5 Layers of the Green Grid. (Source: GANSW and Tyrrellstudios 2016a)

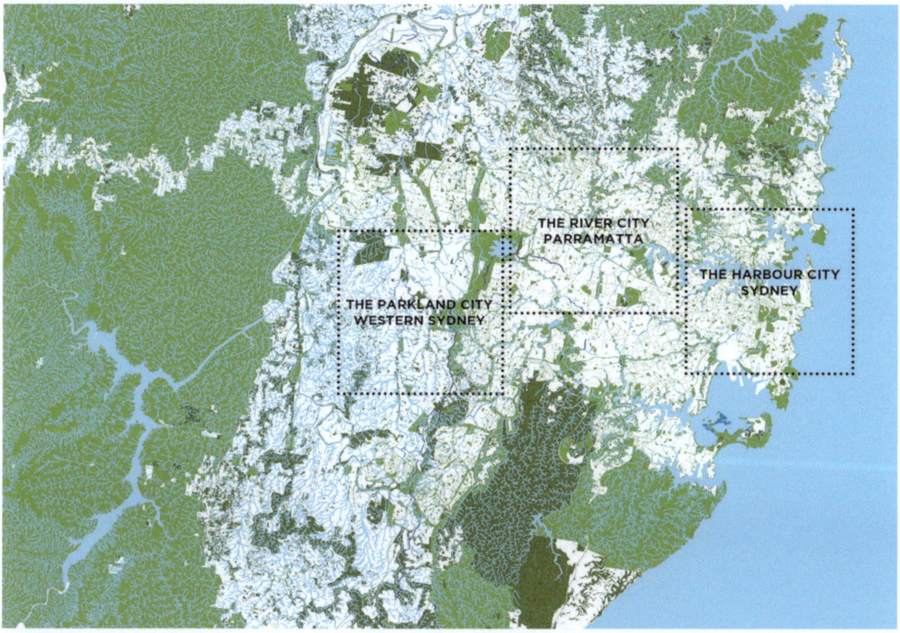

Fig. 8.6 Map of three cities. (Source: GANSW and Tyrrellstudios 2016a)

The Central City district includes the Local Government Areas of Auburn, Blacktown, Cumberland City, Parramatta and The Hills Shire. The Central City district is one of the five districts of metropolitan Sydney as defined by the Greater Sydney Commission's (GSC) District Plans.

The West Central District report (GANSW and Tyrrellstudios 2016b) was one of six reports produced for each district within metropolitan Sydney. The report provides preliminary prioritisation of Green Grid opportunities in terms of their strategic potential for the establishment of a new interconnected high performance green infrastructure network, which will support healthy urban growth.

The landscape setting of the Central City District is comprised of three major water catchments and their associated estuarine and fresh water areas: the Hawkesbury Nepean, the Parramatta River and Sydney Harbour, and the Georges River and Botany Bay catchments (GANSW and Tyrrellstudios 2016b).

Built-up areas are interspersed with industrial estates and feature extensive open space systems along the creek corridors, together with native vegetation. They are complemented by a system of parks, including a number of large regional parks with well-developed sport and recreation infrastructure. They include the Western Sydney Parklands, Parramatta Park, Sydney Olympic Park, Rookwood Cemetery and Prospect Reservoir (Fig. 8.7). Prospect Reservoir is of major importance in the Sydney drinking water supply infrastructure network.

The Central City District highlights Parramatta as a major growth area, being at the heart of the district, forming Sydney's second CBD. The Green Grid for the

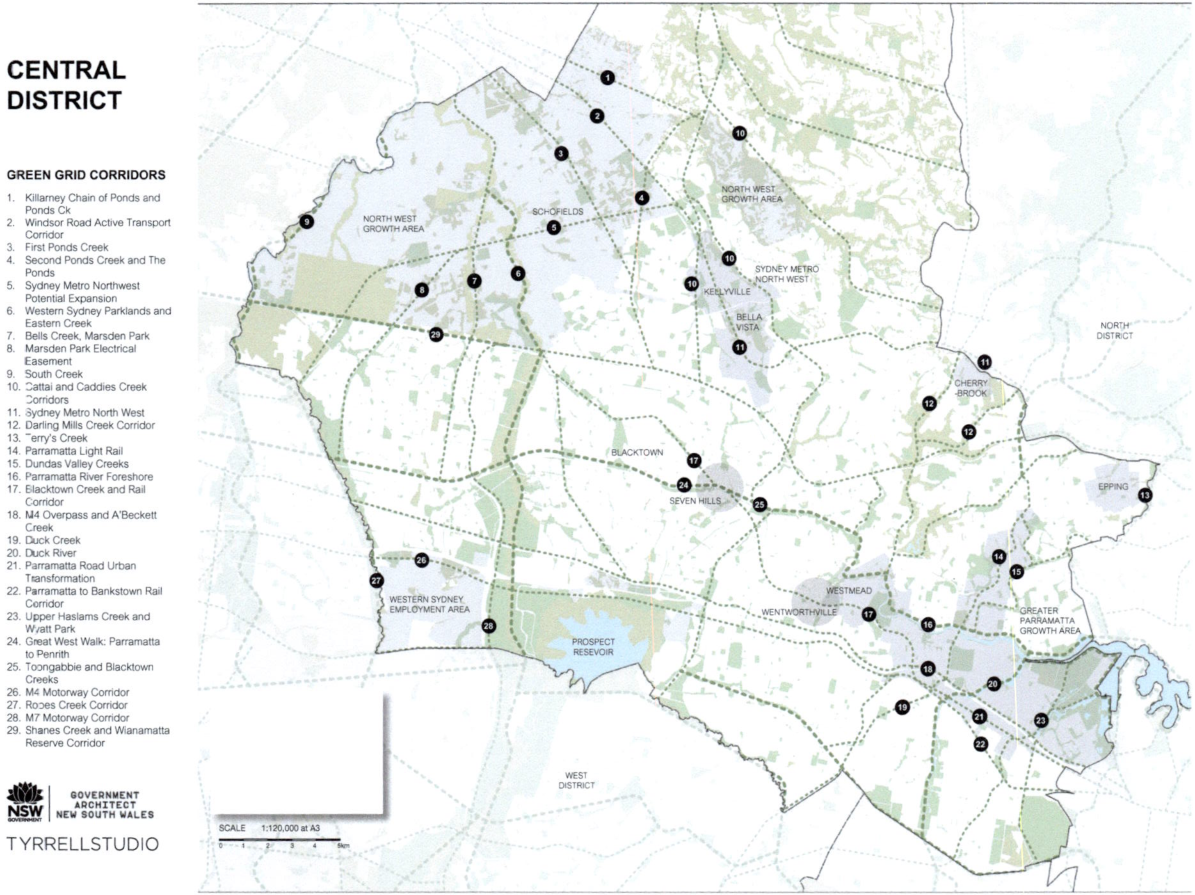

Fig. 8.7 Central city district. (Source: GANSW and Tyrrellstudios 2016b)

Central City District prioritises a number of projects to be implemented. The priority projects represent a mix of project types with hydrological, ecological, recreational, transport and utility infrastructure values, and include:

- Parramatta river foreshore
- Duck river corridor (see local scale case study for more information)
- Ropes creek corridor

8.4.2 *Parramatta Local Government Area*

The City of Parramatta's implementation strategy of the Green Grid is formulated in the document Parramatta Ways (City of Parramatta 2017). Parramatta Ways is a local initiative put in place by the City of Parramatta to deliver Sydney's Green Grid. The council-led project has allowed for specific strategies and targets for the city, with a focus on walkability. The report outlines historical patterns of development that have made walkability a challenge, resulting in broader economic, environmental, health and transport implications. As Parramatta faces future growth there is both an opportunity and a demand to create a green, active and walkable city, which allowed for Parramatta to focus on eight specific objectives for their city:

- Identify and complete priority missing pedestrian network connections
- Increase recreational walking participation
- Increase walk to school rates
- Increase walk to work/public transport trips
- Increase foot traffic to local centres
- Increase tree canopy cover along network
- Increase perceptions of safety associated with walking
- Ensure all residents live within a five-minute walk of the Parramatta Ways network.

Parramatta Ways delivers and expands on the Sydney Green Grid and is aligned to the delivery of the objectives of the Green Grid, as outlined in the District Plan. Prior to the District Plan, Parramatta was used as a pilot project in envisaging the Green Grid project. This initial collaboration with the office of the Government Architect on projects such as the Duck River shows the master-planned potential of Green Grid strategies. Some of these projects, like the Duck River master plan, were precursors to the district plan, which has in turn raised the project's status as a priority project (Fig. 8.8).

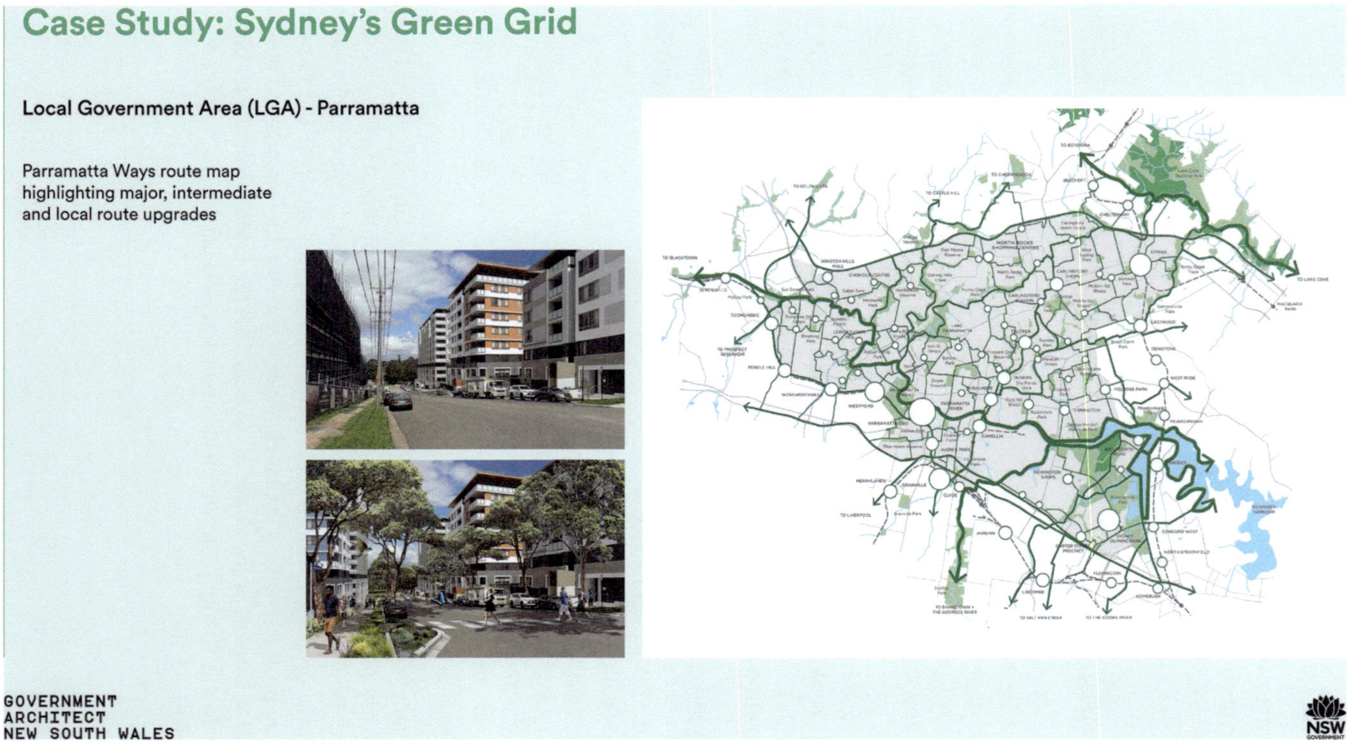

Fig. 8.8 Parramatta Ways, before and after. (Source: City of Parramatta 2017)

8.4.3 Duck River

The Duck River project is an example of a site-specific application of the Green Grid strategy. It is a master plan that improves the public realm though creating a recreational corridor and providing ecological regeneration. The project is part of the Parramatta Green Grid pilot project that implements Sydney's Green Grid strategy. It is also noted as a priority project in the District Plan for the West Central district. The project is located between Auburn and Camellia and is approximately 650 hectares large, 8.5 km in length. The master plan and landscape architecture were delivered through the office of the Government Architect NSW (GANSW) while advice on traffic and transport came from GTA Consulting. The client was the City of Parramatta. The Master plan was completed in 2015 and the implementation is ongoing.

The Duck River Shared Path Master Plan (GANSW 2015) is part of a broader network of high-quality shared pathways envisioned by the City of Parramatta to connect people and places across Western Sydney. These shared pedestrian and cycle ways promote 'active transport' (walking and cycling) for commuting and recreation along the Duck River corridor and throughout Western Sydney more generally.

Duck River is a tributary of the Parramatta River that runs from south to north through a number of local government areas before reaching Sydney Harbour near Camellia. The master plan aims to increase the shared benefits of the waterway's natural landscape features and regional recreation facilities between Parramatta, Auburn and Canterbury-Bankstown local government areas. The initial project brief was confined to the short section of Duck River that runs through the City of Parramatta local government area. Following a competitive tender process, the successful design team encouraged the City of Parramatta to expand the shared path project by partnering with adjacent local councils - the City of Canterbury-Bankstown and Auburn - to realise wider public benefits.

Substantial collaborative work was undertaken by the design team with all three councils to identify connection points for pathways across the river. The success of the master plan lies in the way the design strategy was expanded across local council boundaries and the river itself. The river was understood not as a barrier, but as part of a larger natural and urban landscape that could be used for new access routes between destinations that were historically disconnected.

The implementation of the Duck River Shared Path Master Plan will be led by the City of Parramatta with Auburn and City of Canterbury-Bankstown Council's anticipated to follow. The master plan is being developed into a series of incremental projects to deliver the overarching vision.

The Duck River corridor is one of Western Sydney's hidden natural gems. The master plan unlocks its potential as a place for outdoor recreational activities and shared pathways between neighbourhoods. Through careful landscape design and planned bushland regeneration the project also contributes to the improvement and biodiversity of urban wildlife habitats (Fig. 8.9).

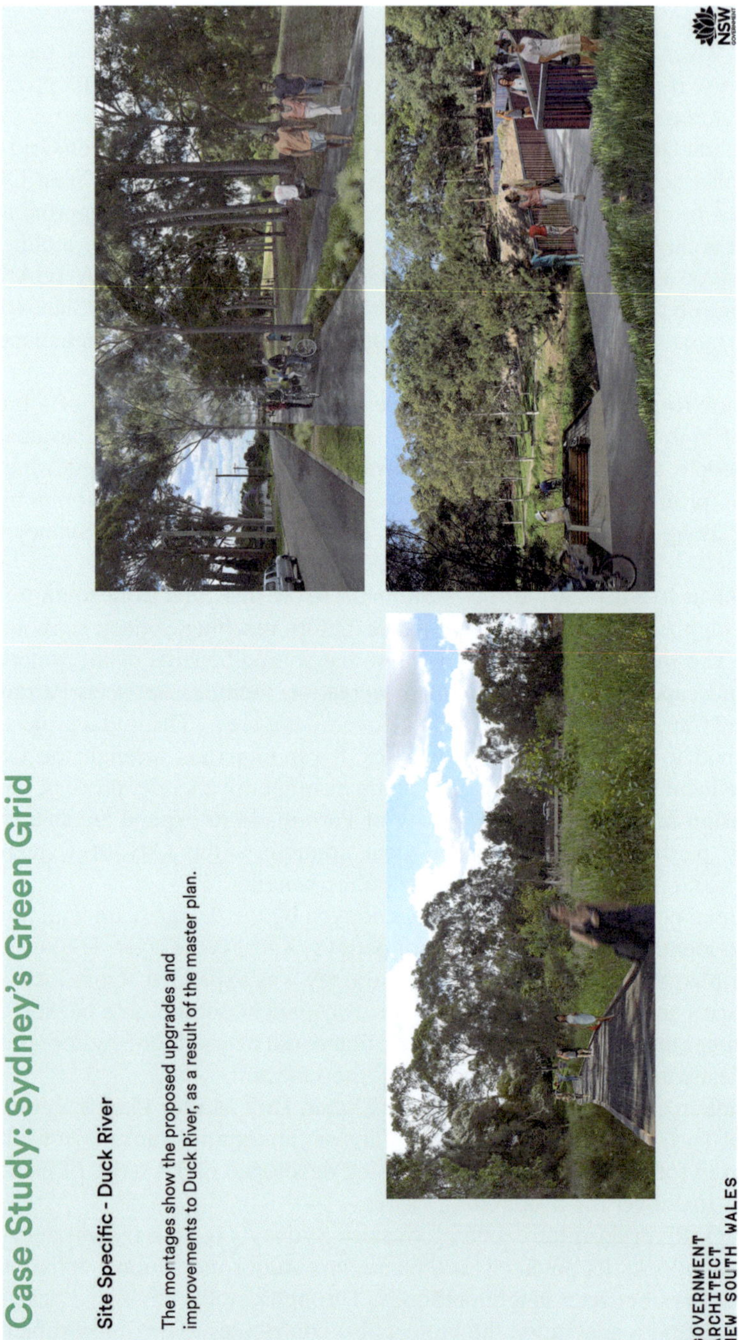

Fig. 8.9 Photomontages Duck river. (Source: GANSW 2015)

8.4.4 Bankstown to Sydenham

The Sydney Metro City and South West extension to Bankstown proposes the upgrade and conversion of the existing 13.4 km rail line to a Metro-rail. This initiative will facilitate faster travel times from Bankstown to the city, improving access to jobs and further enabling housing and employment. The transformation, provides a unique opportunity to revisit the rail corridor space requirements and investigate the opportunity to provide continuous east-west pedestrian and cycle access linking the corridor to the Cooks River and Salt Pan Creek. Reimagining the Sydenham to Bankstown Metro Corridor as an urban linear park provides the opportunity to envisage a wide range of community uses including active recreation, walking, jogging, biking etc.

The relationship between open space and grey infrastructure along this corridor are key to the realisation of enhanced liveability. Improved access to high quality open space underpins the social, environmental and economic benefits that are essential to the healthy functioning of the local built environment.

Investigation of the rail corridor between Sydenham and Bankstown indicates that the rail easements and wide road verges alongside the corridor present a unique opportunity to provide a new open space typology (GANSW 2016). The interweaving of surplus rail land with the adjacent road verges provides a variety of new open space typologies in this highly urbanised setting. For example, active and passive open space, meadows, wetlands, water purification and urban forest.

An open space audit was undertaken for this precinct (GANSW 2016). The audit identified significant gaps in local open space provision. The identified shortfall coupled with an increase in population density and the prohibitive cost of land presents a significant challenge in relation to open space provision in the precinct. The transformation of the rail corridor to metro has identified that not all the easement lands will be required for rail use. This presents a unique opportunity to convert the easement space, which is currently in public ownership into a connected and continuous open space corridor. In addition, the Sydenham to Bankstown rail corridor bisects many of the Sydney Green Grid regional connections, such as the Greenway linking the Cooks River to Iron Cove Bay. This will enable further access to open space walking and cycling routes, as well as opportunities for active and passive recreation and enhanced nature conservation.

The conversion of the corridor to a linear park with walking and cycling provisions provides a unique transport opportunity for local residents and industrial workers. Creative design is called for as the width of the corridor easement varies dramatically. By simply realigning the fence, a range of open space typologies can emerge along the corridor, giving rise to a variety of programmatic opportunities (Figs. 8.10, 8.11, and 8.12).

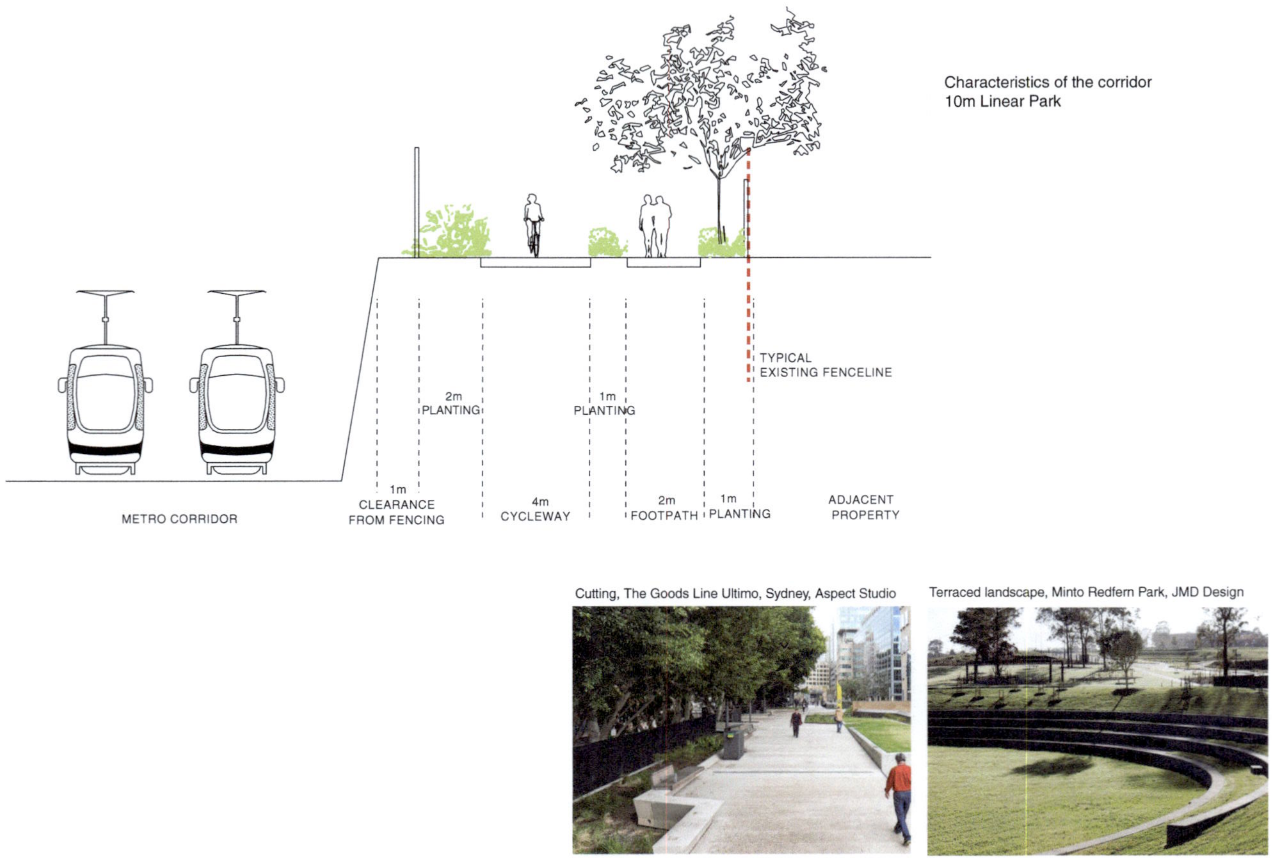

Fig. 8.10 Cutting in Linear Park, Bankstown-Sydenham. (Source: GANSW 2016)

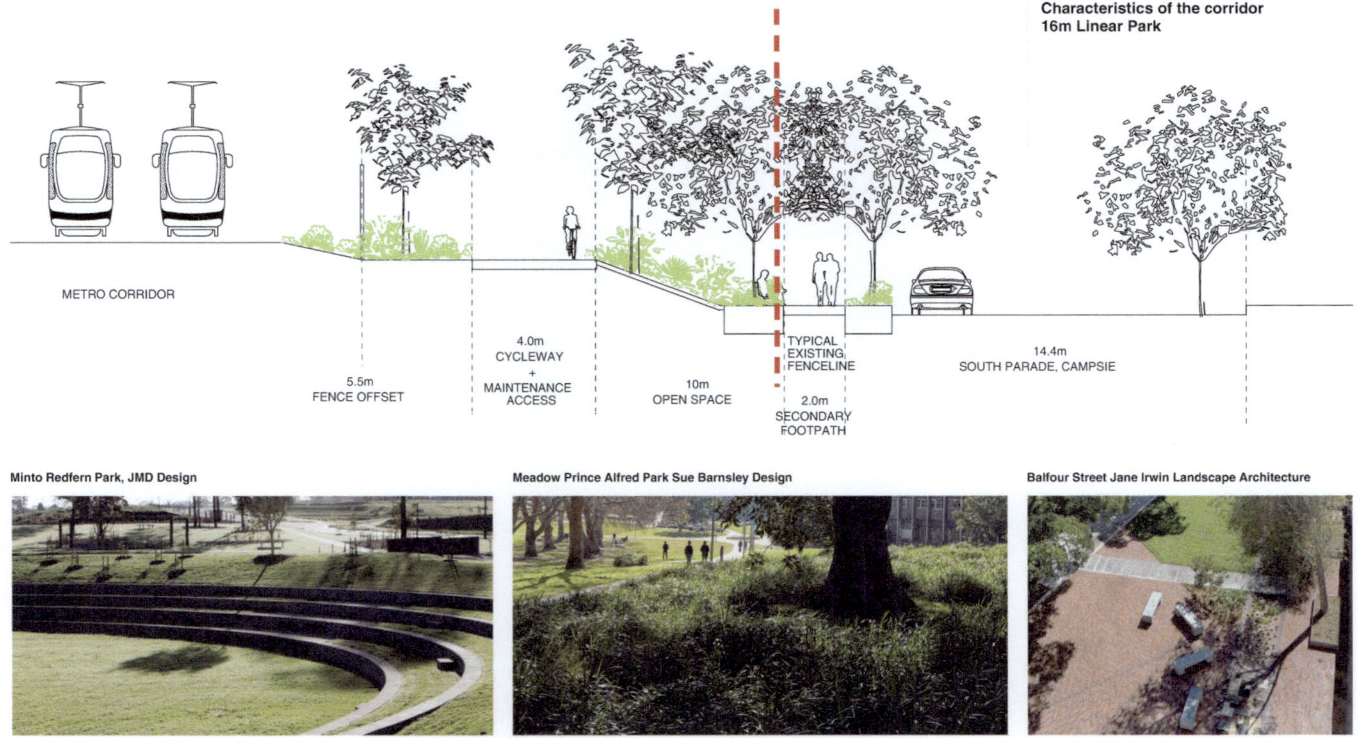

Fig. 8.11 Field and Slope in Linear Park, Bankstown-Sydenham. (Source: GANSW 2016)

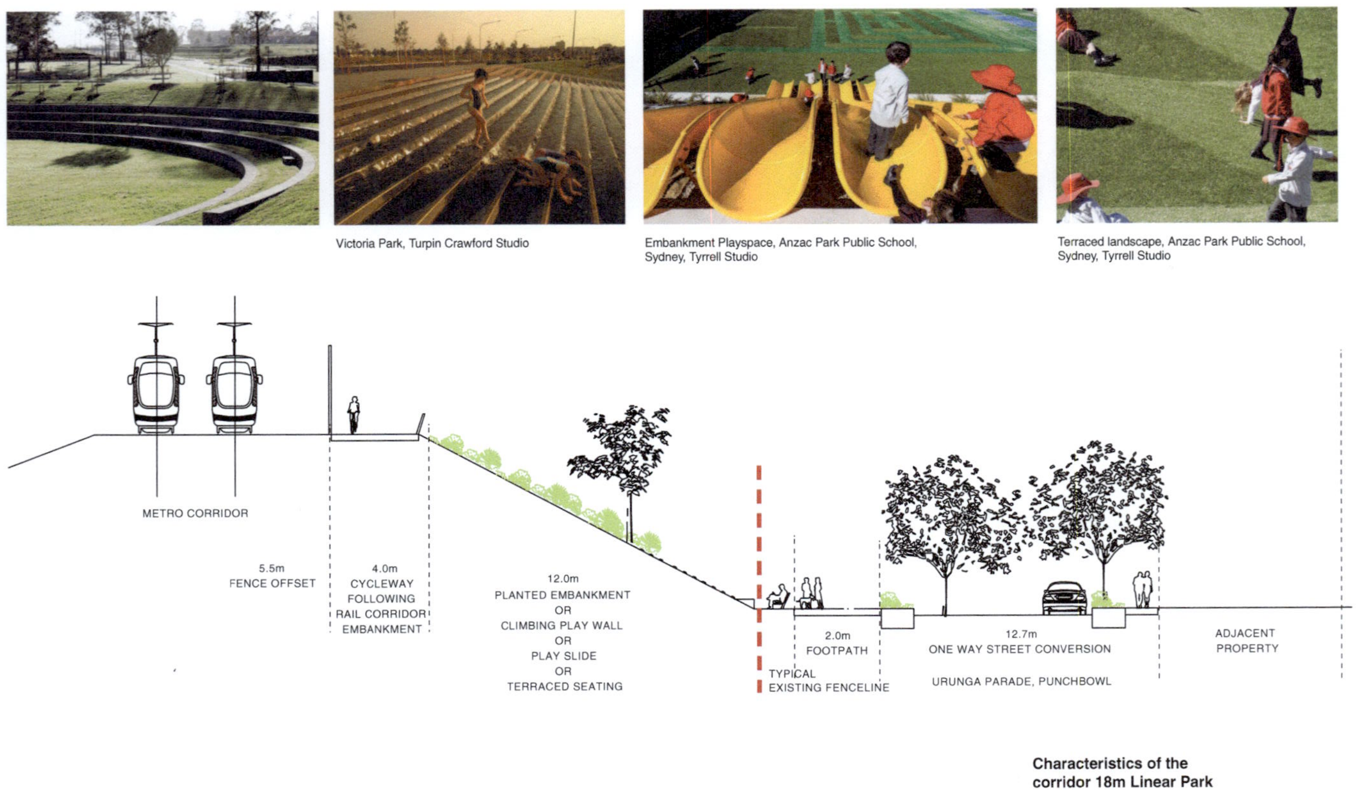

Fig. 8.12 Embankment in Linear Park, Bankstown-Sydenham. (Source: GANSW 2016)

Typologies include:

- Green space - urban forest, native grass fields, meadow and sedge plantings.
- Active space - playgrounds, sports fields, basketball courts, climbing landscapes, dog parks, skate and bike parks and lookouts
- Water Space - bio retention swales, wetlands, water detention basins, water play spaces and reed plantings.
- Community space - community gardens, urban agriculture, urban bee-keeping, flower markets, performance spaces, commercial kiosks, and urban squares
- Active corridor - cycle-ways and walkways.

8.5 Conclusion

The Green Grid envisions green infrastructure as a three-dimensional envelope that surrounds, connects and infuses buildings, streets and utilities. The concept of landscape as green infrastructure provides a potent framework for integrating the work of designers, planners, developers, policy makers and others, and, leveraging this collaboration to achieve larger metropolitan goals.

The awareness of landscape as both a vital resource needing protection and a countervailing force that can be used to positively shape city and subregional development patterns, has seen this initiative included as a key policy directive in the most recent Metropolitan Plan, Towards Sydney 2056, (Greater Sydney Commission 2017a) wherein implementing the Sydney Green Grid is designated a primary action.

Investing in a comprehensive and well-managed Green Grid will provide the living thread that binds sustainable communities together and contributes to the future economic, social and environmental success of our city.

References

Chicago City Plan Commission (2008) Adding green to urban design: A city for us and future generations. https://www.cityofchicago.org/dam/city/depts/zlup/Sustainable_Development/Publications/Green_Urban_Design/GUD_booklet.pdf. Accessed 27 Dec 2017

City of Parramatta (2017) Parramatta ways, implementing Sydney's green grid. City of Parramatta, Parramatta

GANSW (2015) Duck River shared path, Sydney, GANSW

GANSW (2016) Bankstown to Sydenham Linear Park strategy. GANSW, Sydney

GANSW (2017) Greener places, establishing an urban green infrastructure policy for New South Wales. Draft for Consultation. Sydney, State of New South Wales

GANSW & Tyrrellstudios (2016a) Sydney green grid spatial framework and project opportunities. Department of Planning and Environment, Sydney

GANSW & Tyrrellstudios (2016b) West Central district, Sydney green grid spatial framework and project opportunities. Department of Planning and Environment, Sydney

Greater London Authority (2012) Green infrastructure and open environments: the all London green grid. Supplementary planning guidance. https://wwwlondongovuk/sites/default/files/algg_spg_mar2012pdf Accessed 27 Dec 2017
Greater Sydney Commission (2017a) Draft greater Sydney region plan: towards Sydney 2056. State of New South Wales, Sydney. https://www.greater.sydney/draft-greater-sydney-region-plan. Accessed 27 Dec 2017
Greater Sydney Commission (2017b) Draft district plans. State of New South Wales, Sydney. https://www.greater.sydney/district-plans. Accessed 27 Dec 2017
Philadelphia Water Department (2011) Green city, clean waters. City of Philadelphia, Philadelphia. http://www.phillywatersheds.org/doc/GCCW_AmendedJune2011_LOWRES-web.pdf. Accessed 21 Dec 2017

Chapter 9
Supporting the Great Urban Energy Transition

Andy van den Dobbelsteen

Abstract Before discussing the great urban energy transition, which is needed for an even greater global sustainable transition, first a fairy tale.

9.1 Introduction

Before discussing the great urban energy transition, which is needed for an even greater global sustainable transition, first a fairy tale.

9.1.1 The Fairy Tale of Samsø

Once upon a time….

… there was this Danish island called Samsø. It lays in the middle of the Kattegat, positioned between the larger provinces of Denmark, Jylland and Sjælland. Samsø was an old Viking Island that was famous for its potatoes, arguably the best in Europe. Only around 4000 inhabitants lived on the island, spread over about 20 small communities.

Recently, Samsø has become popular and world famous because it officially is the first island in the world that is carbon neutral and energy neutral. But it has not always been like that. In the past the island had an economy of agriculture and national tourism. But after the year 1997 something changed: ever more tourists came to Samsø, not just Danish people yet also foreigners, many of them not even because of the island's beautiful landscape, tasty food and clean beaches, but for serious business: to learn from the island's renewable energy system.

A. van den Dobbelsteen (✉)
Faculty of Design Architecture Building, University of Technology Sydney, Annandale, NSW, Australia
e-mail: a.a.j.f.vandendobbelsteen@tudelft.nl

R. Roggema (ed.), *Contemporary Urban Design Thinking*, Contemporary Urban Design Thinking, https://doi.org/10.1007/978-3-319-91950-8_9

9.1.1.1 How Did It Come So Far?

As should be in a fairy tale, the changes were mainly due to one person, Sir Søren Hermansen, a local entrepreneur who was knighted by the King of the Netherlands, which rarely happens. Sir Søren brought together the island's inhabitants when the Danish government issued a competition for national funding for the best sustainability plan. Sir Søren thought it was ridiculous that Samsø, in the middle of the sea, had to be connected to the mainland for all of its utilities, electricity, cabling, waste water, and so on. So he attracted some experts to work on a proposal to make the island sustainable and asked the islanders to join in. This was a long process, in which he used the old method of a Viking circle, in which everybody has a say and all people are equal (Fig. 9.1). In the end everybody agreed and Sir Søren's proposal won the national competition. With the subsidy and with shareholder money of the islanders themselves they managed to invest in renewable energy measures.

9.1.1.2 What Did They Do?

Sir Søren and the islanders started with making their waste product useful: straw that was normally burnt on the farmlands after summer was now collected and fed into a high-tech oven to produce hot water for district heating on the southern part of the island. For the northern part of Samsø Sir Søren's team invested in a field of solar collector fields that also produced hot water for the northern district heating, in wintertime supported by the burning of wood chips, another waste product from the island (Fig. 9.2).

The biggest investments were made in wind turbines and they ran into trouble just like any other country that wants to invest in wind turbines: people were initially opposed to them until they saw that these wind turbines could be well positioned in the landscape. And there were some farmers who dared to become entrepreneurs in wind energy. In total there were 11 windmills placed on Samsø and 10 offshore. They turned out so profitable that one farmer sold his farm and continues to sell wind electricity.

Fig. 9.1 Sir Søren Hermansen in the middle of a Viking circle in the Energy Academy of Samsø, explaining his plans. (photo by the author)

Fig. 9.2 Straw bales for the southern heat grid (left) and solar heat collectors, wood-chip furnace and hot water storage (right) for the northern heat grid of Samsø. (Photos by the author)

9.1.1.3 The Result

Within 8 years' time the island managed to become energy positive. So Samsø produces more than 100% of its own energy consumption from renewables. It is carbon positive too. Next to the renewable energy this is due to new ways of agriculture that can absorb more quantities of carbon in the soil. Samsø is not fossil free yet because the islanders still use diesel and petrol for their cars and petrol for the ferry boats. Nonetheless, Sir Søren and his team are now working on a trajectory to make the island fossil free. The municipality has purchased electric cars for their staff, charged by a large PV parking roof, and the island has invested in a new ferry run on liquefied natural gas, which in a few years' time can be replaced by biogas from Samsø itself. For the first time, the islanders have to look at their buildings' energy-efficiency to make these also fully fossil free. Creating a circular, organic economy is now one of the key points.

9.1.1.4 It Is Always About People

Regardless of the fact that Samsø is an island with quite some space and a limited number of inhabitants – however with great numbers of tourists in summertime – its technical accomplishment speaks for itself and should be applauded because so far no municipality has achieved this. Most important of this real-life fairy tale, however, is that the energy transition caused a great social coherence among the islanders and big local participation in the organisation of the Samsø's renewable energy system. People are really proud of what they have established and the transition to fossil freedom encounters hardly any opposition anymore. The fairy tale has not ended yet, but I am pretty sure that before 2030 Samsø will be fossil free, yet again an international example for many communities in the world, with its people, as a fairy tale should end, living happily ever after.

The unfinished story of Samsø demonstrates that seemingly unattainable goals can be achieved, giving hope to the world. As Sir Søren Hermansen says himself: technology is not the problem, money is not the problem, but the problem is the will of the people.

9.1.2 Definitions and Scopes

9.1.2.1 Tiered Ambitions

When discussing carbon neutrality, there is a tier of ambition steps that one can take, based on clean and clear definitions of the ambitions that so often are misused and misunderstood. The following text therefore not necessarily agrees with other definitions used by scholars. They are based on two studies: the Lidl Nederland sustainability research (Graamans and Dobbelsteen 2017) and the City-zen vision document for the Amsterdam Roadmap (van den Dobbelsteen 2017).

- Carbon neutral is actually the easiest ambition level one can reach: having net zero carbon emissions over a year's time. Once carbon neutral one does not have to be energy neutral though, because carbon emissions can be compensated for. For instance, when you purchase a flight ticket, with most air companies you can pay extra for your ticket so that your carbon emissions are being compensated for. This extra money can go to forestation projects or are used for carbon trading, which at its core is a fake sustainable solution. So, when you cannot solve all energy issues and there still are carbon emissions, compensation can still make you carbon neutral.
- The next ambition tier is energy neutral. This means that over a year's time one produces as much renewable energy as one uses. This means that renewable energy can also compensate for fossil fuel consumption. So one is still allowed to use fossil fuels in an energy neutral system. See Samsø at present.
- Therefore, the next tier is to become fossil free, just as Samsø's next ambition, and then no fossil fuels are allowed anywhere in the system, no natural gas, no mineral oil, no coal. Assuming no nuclear power, a fossil free system is also circular for its energy use, but not yet for other flows, such as water, materials and food.
- This means that the next tier in ambition is becoming circular with all resources: energy, water, materials and nutrients. A circular economy (Ellen MacArthur Foundation, based on Cradle to Cradle (McDonough and Braungart 2002) has two possible cycles it can refer to: the technical cycle, focused on high-quality technical products that can be endlessly reused, recycled or reprocessed, and the natural cycle, which is based on replenishment of renewable sources and entails safe, non-toxic wastes to nature.
- The highest ambition one can have in this list is becoming regenerative. From the moment a system is circular, all resources are reused and recycled or replaced by

renewables; one could call this a self-sufficient, autarkic system. Understanding that at present the earth is already overstretched, unbalanced and damaged by human developments, a truly sustainable situation would arise when a system also repairs old damages, restores natural reserves and rebuilds capacity that got lost in the past. This is a regenerative system.

Being listed first in the ranking of ambitions does not mean that becoming carbon neutral is an easy goal for cities, by the way.

9.1.2.2 Sources of Energy Use

Mostly, energy neutrality only refers to building-related energy, sometimes also user-related energy, but it is seldom connected to the embodied energy in building materials.

Next to this hardware of a building and equipment, living is fairly important. How do we live? What do we eat? Where do we travel, and by which mode of transport? What products do we buy? And what are the implications of these for the use of energy and the production of carbon? One could investigate one's own life and see the influence on carbon emissions of living (van den Dobbelsteen et al. 2017).

So far, the things mentioned relate to one's personal life. However, there also is a substantial share of energy used by elements outside ourselves, which we utilise but are not directly responsible for. Think of infrastructures, utilities and facilities, industries, public buildings and commercial objects. No one feels personally responsible for these elements of the public domain that also cause environmental damage, but we are all part of the problem when we use them.

This shows that it is very difficult to solve the energy and carbon issue within cities by one handclap. Individually, you probably have to start closest to home, with the things easiest to commence with.

9.1.3 The New Reality

Cities have to deal with three boundary conditions that are different than they have been in the past 200 years. These are: climate change, and depletion of fossil fuels, scarcity of resources. Climate change is already happening and we had better be prepared for extreme weather events. Nevertheless, for future generations we should avoid even harsher climate extremes and become fossil free. Not least, we need to start using only materials that fit a circular economy. Spatial planners, urban designers, architects, building engineers and consultants should be aware that from now on they have to deal with these three issues that their predecessors never had problems with in the past, the decadent age of abundance, globalisation and climatic stability.

So it is necessary to become climate adaptive, fossil free and circular. The following section will discuss how this can be achieved.

9.2 Solutions

9.2.1 *Climate Adaptivity*

Not really related to climate change but very important for the human climate is Fig. 9.3, showing that death has a great partner in air pollution. Many parts in the world are actually not very healthy to live. In Europe, the life expectancy reduction in months, of people, is devastating in some regions. If all EU measures to abate fine dust particles would be taken, differences in life expectancy are still big. And many EU member states oppose to taking these measures, mainly for economic reasons. This could be qualified as a criminal act, because it puts the lives of people at stake.

When talking about greenhouse gases and the impact of carbon dioxide on our climate, the International Panel on Climate Change (IPCC 2014) is unprecedentedly clear: 97% of all peer-reviewed scientific publications agree on climate change and the human influence on it. So, as a non-climatologist one should not dare to doubt this authoritative group, and so should other laymen. To underpin their authoritative work, everything predicted about the climate in the 1990s has already come true in recent years.

Experts already state that staying within the two-degrees global temperature increase will probably not be attainable, implying serious shifts in climatic conditions. This means that certain plants and animals will not be able to adapt fast enough. This is called extinction. And of course climate change will have devastating effects on sea level rise, storms and excessive rainfall, opposed to serious heat and droughts elsewhere, or in fact in the same country. A few years ago, in 1 week's time, Australia experienced a new extreme temperature level in the outback and floods in Queensland that had not happened for centuries. With the ever-increasing

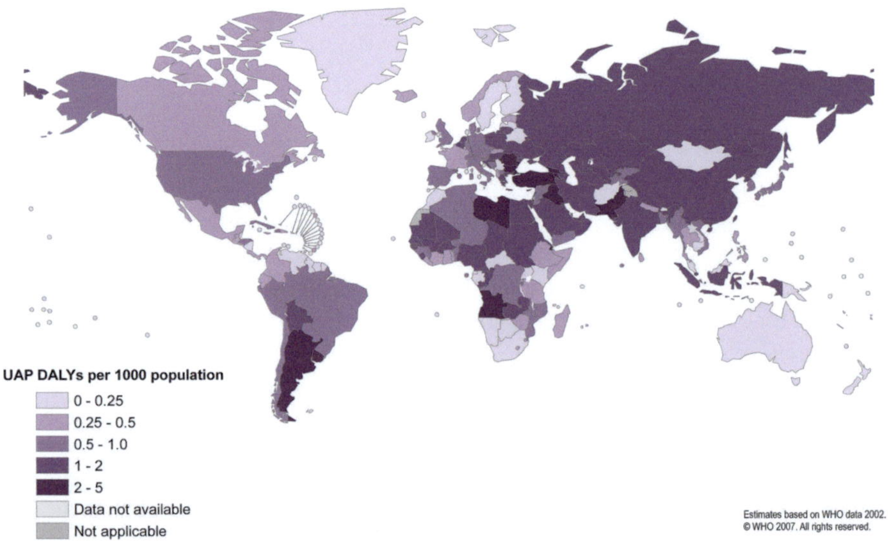

Fig. 9.3 Urban air pollution DALYs (Disabled Life Years) (WHO 2007)

frequency and intensity of hurricanes and tropical storms we may conclude climate change is already here.

What to do about climate change when it is already happening? We would better be prepared and adapt to the inevitable changes.

9.2.1.1 Water

In The Netherlands around 40% of land surface is below sea level, and increasingly so with sea level rise and land subsidence due to oxidation of peaty areas. In order to prepare for excessive rains as in Texas, August 2017, many of the country's polders should perhaps be flooded again to create more buffer volume for water. Closer to home, functional domestic rainwater storage, such as is promoted in Australia and Belgium, to be used for toilet flushing, gardening or car washing, helps retention of storm water and to bridge water supply in drier periods. Further than this, offices like Atelier Dreiseitl (Germany), opMAAT and NIBE (Netherlands) were designing integrated water systems fully reliant on rainwater and water purification already in the 1990s.

When a flood cannot be averted, we can still approach the water problem positively by creating neighbourhoods and houses that are flood safe. There are many possibilities to do this, as the Zuidplaspolder study explored for a Dutch polder that is threatened most likely by a 1.3 flood (Fig. 9.4).

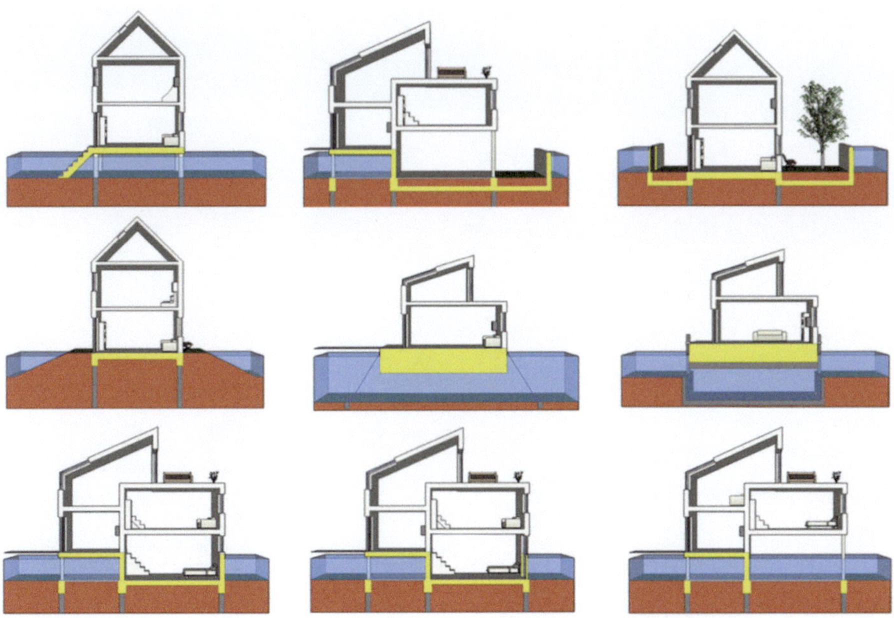

Fig. 9.4 Flood-proof dwellings for the Zuidplaspolder in The Netherlands (van den Dobbelsteen et al. 2008)

9.2.1.2 Heat

Heat is another important factor of climate change, in areas prone to drought and in cities, where absorption of the sun leads to an enormous temperature rise in cities, aggravated by anthropogenic heat coming from vehicles or air conditioners, and not compensated enough through lack of green or flowing water (Fig. 9.5). Statistically, there is an increase in mortality when night temperatures do not decrease enough anymore, and this is exactly what is happening in our cities. Meanwhile, comfort and energy performance requirements for cold and temperate climates are still mainly focused on heating in wintertime, whereas cold demands become ever more significant. Building regulations do not cater for this change yet.

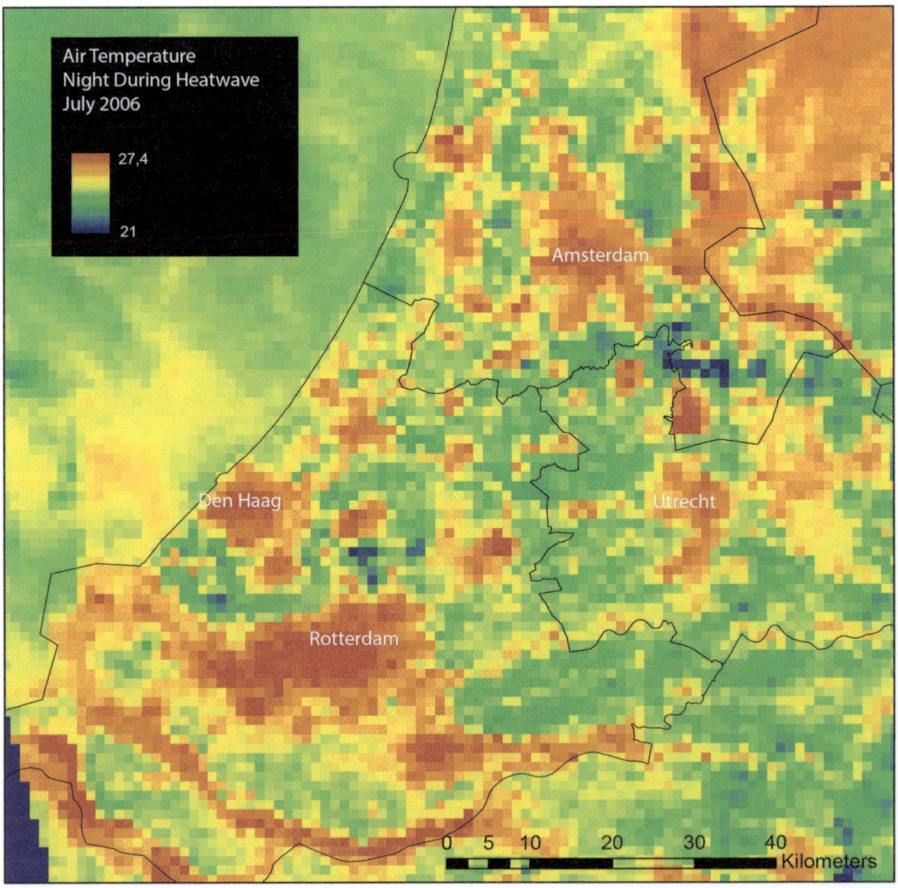

Fig. 9.5 Satellite image of the western part of the Netherlands during the 2006 heatwave, indicating temperature differences up to 7° between cities and their rural surroundings (van den Hoeven et al. 2014)

People from relatively cool climates could learn from regions that are warmer and look at the vernacular examples therefrom, because these basically show the way the people solved it back in time when they did not have all means at disposal and therefore had to be smart about the design of their buildings and communities.

For currently still cool climates switching to heat pump systems is a solution, since these can cool and can heat in an energy-efficient way. The source of these heat pumps should be the soil or open water, because air would cause a temperature increase in the city. Furthermore, applying photovoltaic panels and solar heat collectors take away energy that would otherwise heat up the city. Tarmac roads and stony pavements could also become solar heat collectors, taking away undesired heat and producing hot water that can be used directly or stored for a colder season.

9.2.1.3 Adaptation Meets Mitigation

Therefore, we can find solutions that serve both heat mitigation and energy production. We need to create climate-robust areas by introducing much more green, introduce flowing water, and adapting the houses to a warmer climate, for instance. Particularly novel are solutions that help to acclimatise cities better, comparable to how we design our buildings' indoor climate. Examples of this are:

- Plants are natural air-conditioners, so applying green infrastructures throughout the city helps to cool it down (e.g. Klemm et al. 2015).
- Wisely chosen surface materials, taking into account albedo, or reflectivity, helps to avoid absorption of solar heat. White roofs reflect the sun to the sky, but not all surfaces need to be light coloured: white, glimmering facades reflect the sun to other surfaces such as pavement, which will then heat up, while darker facades can create a thermal updraft, so that the heat is actually released into the air above (e.g. Taleghani et al. 2014).
- Deliberately creating thermal drafts via dark-coloured wall and solar chimneys will release heat into the air above and draw in fresh air from elsewhere, which might be cool spots such as parks (e.g. Kleerekoper et al. 2015).
- The most technical solution is to actively cool public squares or sidewalks. This sounds like using a lot of energy but another way to put it is: collecting solar heat and using that elsewhere or some other time. So this is when climate adaptation meets climate mitigation, reducing carbon emissions.

9.2.2 Energy Neutrality

Cities cannot function without energy. Energy is very important; in our modern society it is needed for everything. If cities experience a serious black-out, except for buildings that have a backup system of diesel generators, such as hospitals, everything basically stops to function. If we want resilient cities, these will have to take better care of themselves, not becoming autarkic as such but be more resilient, in the sense that they can withstand these problems, at least for a certain time.

It starts with making buildings more energy efficient, using the New Stepped Strategy (Fig. 9.6). First step (0): understanding really well what to design for, what are the local conditions, the climate, the underground, the circumstances, before you try to reduce the demands (step 1), with urban and architectural design measures that do not require any technology yet. Next step (2): think smartly about reusing waste flows, because if not, it will heat up the city and there will be a loss of useful energy. The last step (3) is to produce the remaining energy demand from renewables. The order of these steps can vary, because sometimes producing renewables can be easier and cheaper than for instance saving energy; this depends on the situation, which is why the circumstances need to be studied first.

This strategy was expanded to different urban scales in the Rotterdam Energy Approach and Planning (REAP) (Tillie et al. 2009), specially promoting exchange of energy (step 2) on the neighbourhood, district and city level.

Figure 9.7 is an image of heat demand and heat potentials of the city of Rotterdam. One can see that the heat demands (hollow blue cores) are too high to be solved by renewable sources such as solar (yellow) and geothermal (red). Next to these there are different distributed sources of waste heat, which could be small industries, supermarkets, or other urban functions with excessive heat. This heat can be

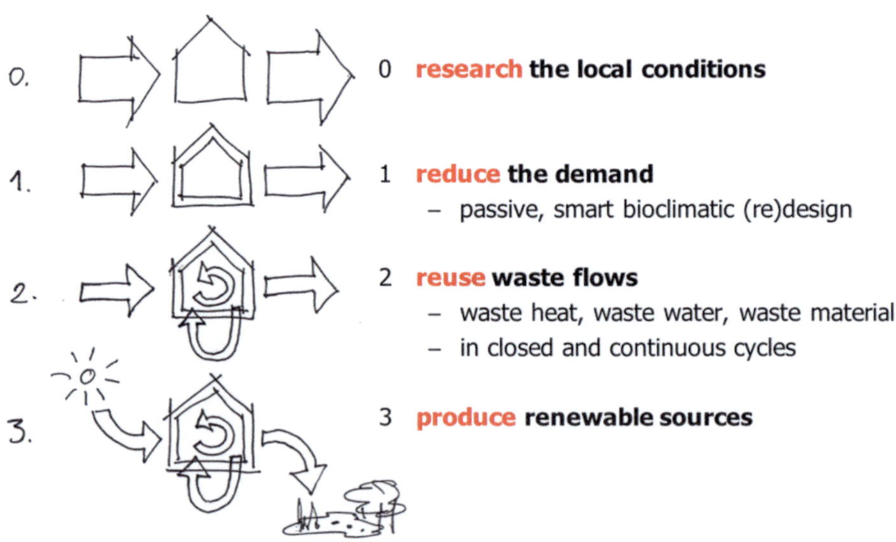

Fig. 9.6 The New Stepped Strategy: research, reduce, reuse, recycle (adaptation of van den Dobbelsteen 2008)

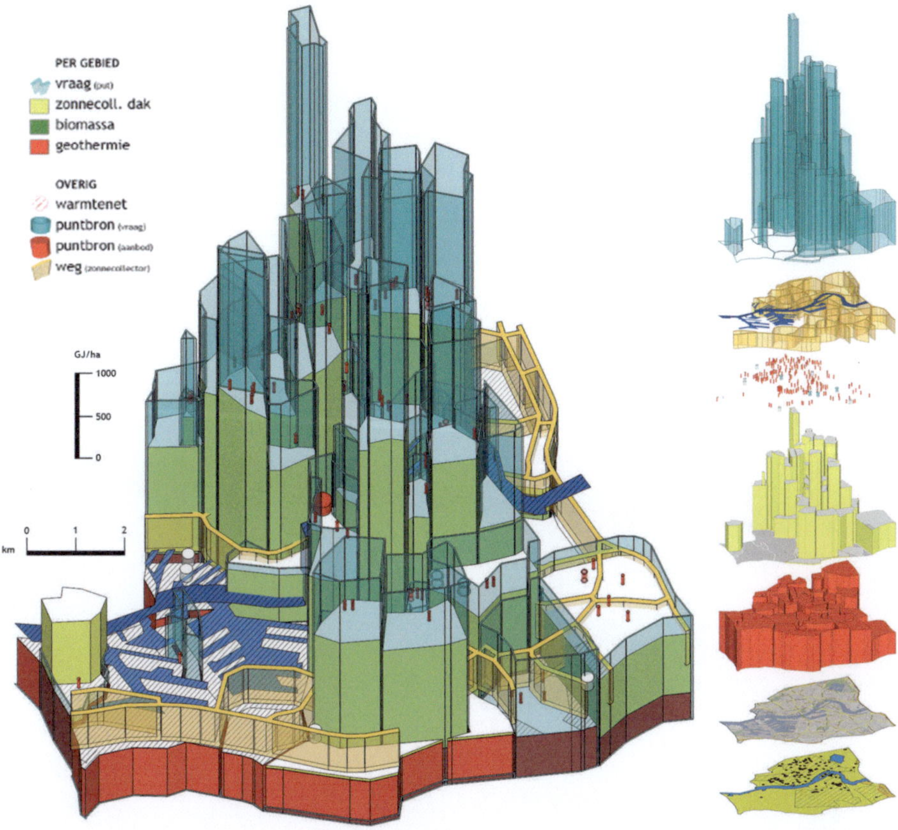

Fig. 9.7 3D heat map of Rotterdam. (Broersma et al. 2013)

exchanged with other functions. Supermarkets, for instance, require cooling all year around, emitting waste heat into the air continuously, while they are usually built next to their customers in houses that require a lot of heat. These buildings could have one energy system based on a heat pump.

So there are different sorts of urban heat potentials that are currently not used, and if the city does not, they will heat up the city, that is a simple equation.

Energy potentials are different everywhere, surprising sometimes. In Flanders, Belgium, there are old coalmines that are not in use anymore. These mines are very interesting because they are an open infrastructure in the underground through which geothermal heat of a certain temperature can be reaped (Fig. 9.8). This geothermal heat can be coupled to settlements at the surface, buildings that are already there or new developments, even thermal lakes if one desires so.

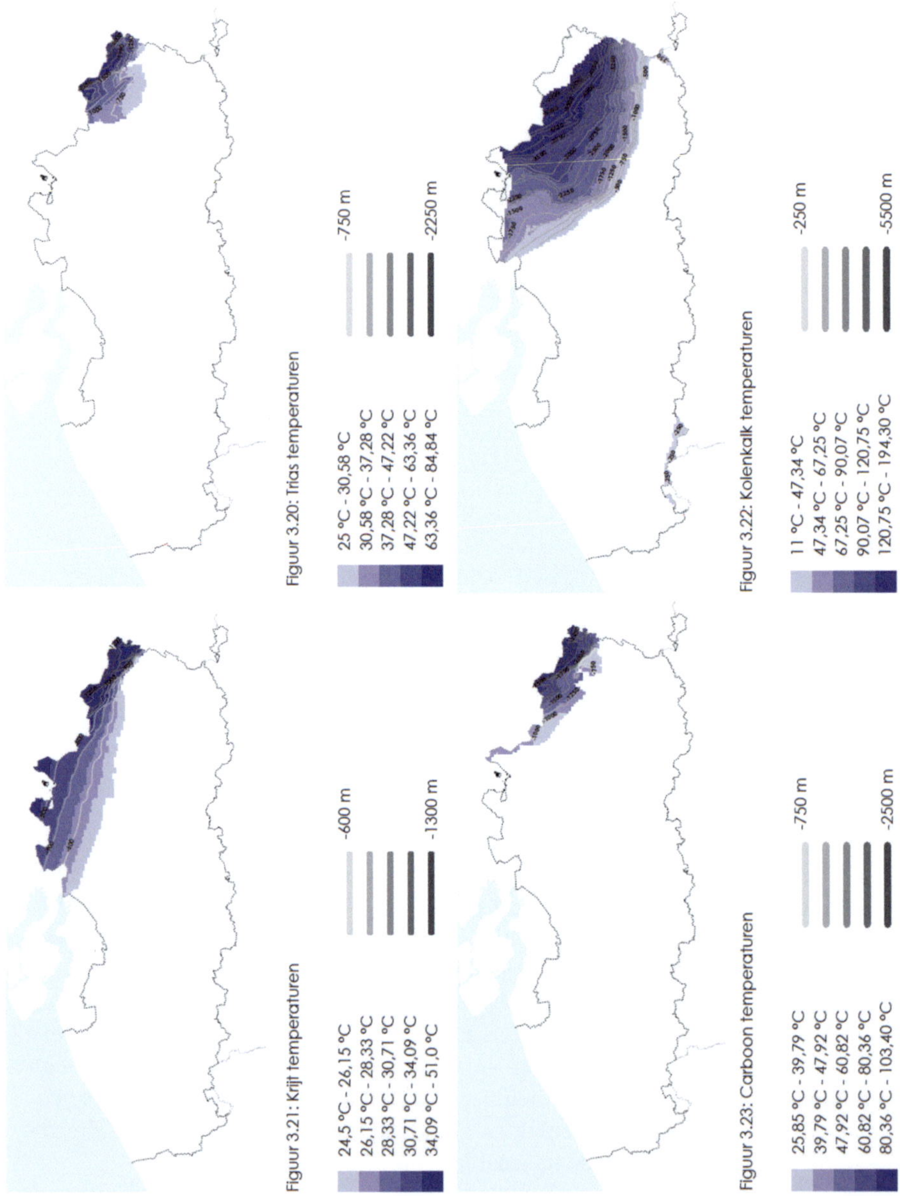

Fig. 9.8 Geothermal potential in Flanders, Belgium. (Labo Ruimte et al. 2016)

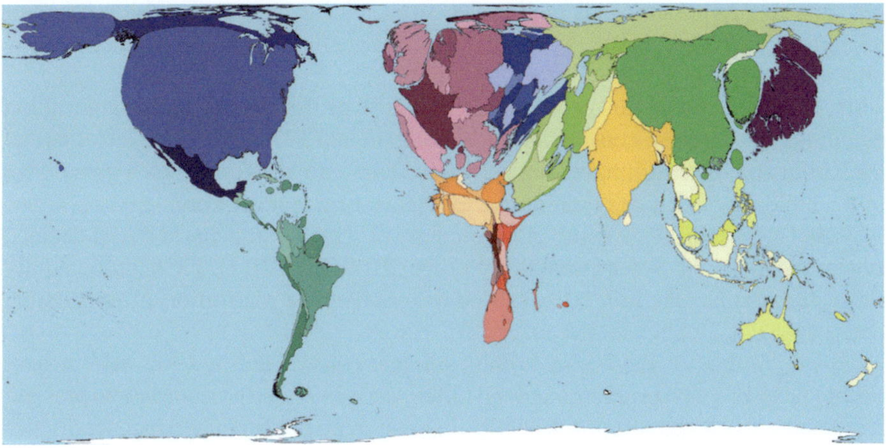

Fig. 9.9 The world as it would be if all countries were as big as the space they need for their resources, water, energy, food (www.worldmapper.org)

9.2.3 *Circularity*

There are a lot of resources going through the building industry at the moment, causing depletion and environmental damage. And there still is a lot of inequality when we see Fig. 9.9. The Northern Hemisphere is blown up compared to the South, and if we would not begrudge Africa and South America to gain the same level of prosperity as we have, the earth would simply have to blow up. She is already overstretched.

So we have to do something to deflate the Northern Hemisphere. A circular economy might be the solution for that. For circularity, cities need to become intelligent organisms (van den Dobbelsteen et al. 2010). We can learn a lot from ecology and from bio-mimetics and we should translate these to cities. At present, cities are basically a collection of separate buildings that have to be connected to all sorts of utilities; there is no intelligence in the interaction.

9.3 Dealing with the Existing Building Stock

9.3.1 *The Real Challenge*

If the ambition is to make effective steps towards the future, then the numbers, both in quantities and improvement potential need to be looked at. For cities this means the greater urban areas and predominant buildings that currently form a sustainability problem. In Europe, the terraced house is a widely used residential building type. The ones built before 1975 are commonly energetically very inefficient, which is not only environmentally damaging yet also causing high energy bills for their inhabitants. With rising energy prices people sometimes cannot pay their energy bills anymore.

In The Netherlands this applies to around 20% of the housing stock, 1.4 million dwellings, which were constructed between 1946 and 1975 and which lack thermal insulation and often still have single glazing. These houses form such a great share of the building stock that demolition and replacement is no option.

When at the Delft University of Technology (TU Delft) students and staff decided to participate in the Solar Decathlon Europe 2014 (SDE2014), the team opted not for a flashy solar pavilion but work on the sustainable renovation of an existing house (van den Dobbelsteen et al. 2015).

The team took a real house as the example, constructed in 1960, with a poor energy label E, moist problems and too little space for a family, according to present-day standards. Nevertheless, the people living in these houses experience feelings of tenderness about these houses, because they grew up in them and the dwellings developed a sort of spirit that new-built houses do not have yet. So how could their home be preserved but the house improved, in such a way that its household did not have to move to a hotel. This is also why the project is called 'Prêt-à-Loger', ready to inhabit.

9.3.2 *Redesign of the Original House*

The first idea was to drape a new sustainable skin over the building ('Home with a Skin'), to solve all the problems by sustainable technology on the outside. Before elaborating this idea, the student team studied the local circumstances first.

9.3.2.1 Climatic Conditions

The average temperature in The Netherlands, also in the region of the original house, is around 10 or 11°, still pretty cool, but good in summertime for cooling and in wintertime to pre-heat even colder air coming in through soil ducts.

The case region receives around 850 millimetres of precipitation. We calculated that the dwelling's roof surface area is sufficient to provide enough rainwater for toilet flushing and for the garden.

There is quite a lot of wind in the coastal region of The Netherlands, an average wind speed of 5–6 m/s, but for reasons of risks it is difficult to put large wind turbines because of the vast Dutch green horticulture area over there, and small urban turbines are still relatively inefficient.

Regarding the sun, in The Netherlands a horizontal plane receives one thousand kilowatt hours per square meter per annum. Our particular house has its garden to the South East and the street to the North-West. So for solar energy the focus lies on the garden side.

9.3.2.2 Applying the New Stepped Strategy

First step regarding the original house was to reduce the energy demand by adding thermal insulation, replacing the windows with high-performance low-emittance glass, and adding Solatubes on the roof for daylight access into the home.

Next the soil would be used for pre-cooling and pre-heating of fresh incoming air, before it goes through a heat exchanger to create maximum efficiency. However, in Versailles, France, where the SDE2014 competition was to be held, it was not allowed to dig into the holy of ground of Louis Quatorze, the French sun-king, so a different solution had to be found and phase-change materials (PCMs) were introduced in the crawl space, which could pre-cool the air before it comes into the house.

The biggest intervention was introducing a greenhouse, the glasshouse to the South Eastern elevation, on which photovoltaic cells could be applied. The most important, however, and smartest part of this element, is its use for collecting solar heat. An adiabatic heat exchanger in the glasshouse notch, underneath the PV cells, produces hot water from the warm air that is collected there. A heat pump boosts the water temperature to 55°, enough for hot water purposes and for the central heating system.

Furthermore rainwater is collected from the roof in a tank underneath the greenhouse and is used for toilet flushing and watering the plants.

Last and not least, the project demonstrates to the dwellers of these houses how they can create a greener living environment, produce their own food, and use plants that fit the local climate, so that these can be maintained without pesticides or toxic means.

Figure 9.10 shows a cross-section of the Prêt-à-Loger renovation plan. In wintertime, the greenhouse is a heat buffering space. In summertime it can be opened completely so becomes a terrace, a part of the garden. In the in-between spring and

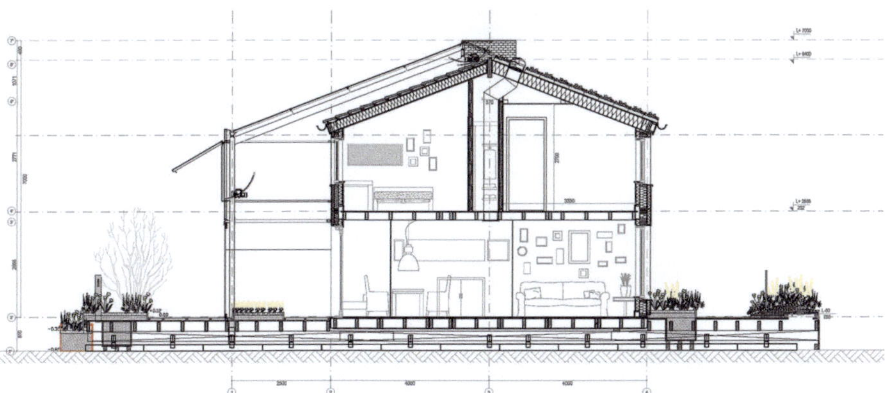

Fig. 9.10 Cross-section of the renovation plan (TU Delft Prêt-à-Loger team 2014)

Fig. 9.11 The Prêt-à-Loger house at the Solar Decathlon Europe 2014 competition in Versailles, France. (Photo by the author)

fall seasons the greenhouse has a pleasant temperature and can be used for working or living.

Figure 9.11 shows what the Prêt-à-Loger house looked like during the SDE2014 competition. The cut through the neighbours' home was used to illustrate the difference in cross-section before and after the intervention.

9.3.2.3 Success

The TU Delft SDE2014 team won five prizes, among which the first prize for 'Sustainability' and for 'Communication & Social Awareness', but as the only renovation entry it won also the second prizes for 'Energy Efficiency' and 'Safety & Construction Management'. The house was dismantled and reconstructed at the TU Delft campus, where it until now has functioned as a visitable and inhabitable demonstration object, receiving a lot of visitors.

9.4 Working on the Urban Energy Transition

9.4.1 The City–zen Roadshow

Within the EU FP7 project called City-zen so called Roadshows are undertaken throughout Europe to help cities in their energy transition towards renewables. With the City-zen Roadshow a group of researchers from different European universities and institutes visit a city and do a one-week intensive workshop with local stakeholders, following The SWAT Studio, an explorative two-weeks student workshop from TU Delft.

The City-zen Roadshow methodology is a thoroughly organised week schedule including lectures, field visits and stakeholder interviews. However, the largest part of the work is done in interactive sessions between the researchers and stakeholders, creative work informed by hard data from the city and validated by energy and carbon calculations. The Roadshow team analyses the use of energy and emission of carbon dioxide, in total numbers and per household (Fig. 9.12). These carbon data can be coupled to the forest area one needs to plant to compensate for the emissions, but the eventual goal is to avoid this principle of compensation. The designs, calculations and proposed embedment in governance eventually leads to a city vision on different scales – city, neighbourhood and building level – which is presented to decision makers of the city on the final morning of the Roadshow.

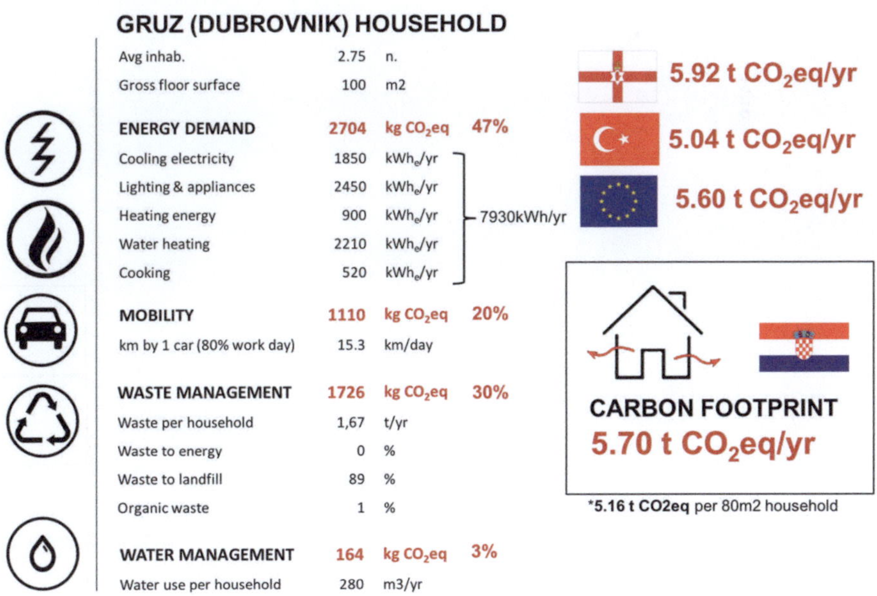

Fig. 9.12 Energy consumption and carbon emission calculation for Gruž, Dubrovnik (Croatia), compared to earlier City-zen Roadshow calculations for Belfast (Northern Ireland) and Izmir (Turkey) (City-zen Roadshow team 2016; table by Riccardo Pulselli, University of Siena)

9.4.1.1 Dubrovnik

In fall 2016 the City-zen Roadshow visited Dubrovnik, Croatia. Most people know this city for its beautiful historical centre – home-base for the imaginary city called King's Landing in the popular series Game of Thrones – but Gruž, the port town of Dubrovnik is a relatively poor, energy-inefficient town. The compensation for its energy consumption would be 1100 hectares of forestland (see Fig. 9.13). The goal was to come up with a plan for a carbon-neutral Gruž, meanwhile serving to full energy transition of Dubrovnik.

The climate of Dubrovnik seems quite favourable: 16 °C on average, meaning the soil has a useful temperature suited for cooling and pre-heating; more than 1000 mm of rainfall, so that in fact all water consumption could be served with that. The solar potential is great, wind turbines would be very effective on the coastal ridges, and the sea alongside can be used as source for heat and cold.

Seeing these potentials few cities have, the problems the city was dealing with in summertime, tens of thousands of tourists visit the city daily, suffocating a normal life, and many of these tourists arrive by cruise ship, the dirtiest floating towns around. Gruž suffers severely from the fumes and soot emitted by the cruise ships.

Fig. 9.13 Gruž (the area depicted in yellow) and its carbon sequestration area as forestland (in green) to compensate for the energy consumed in this part of Dubrovnik (City-zen Roadshow team 2016; image by Riccardo Pulselli, University of Siena)

9.4.1.2 Solutions

Understanding the real negative issues, the cruise ship tourist issue was tackled first. Requesting these cruise ships to bring their waste water to Gruž, instead of accepting to release it into the sea, as is common, waste water treatment by algae enables the production of biodiesel and nutrients that are used to grow food (Keeffe 2009), both of which can be sold back to the cruise ships. This green port solution (Fig. 9.14) would create a double win for the local community of Gruž.

For the buildings of Gruž an energy renovation scheme is proposed, starting with the larger apartment blocks, which could be turned into energy neutral with post-insulation and PV, algae or greenhouse facades (Fig. 9.15).

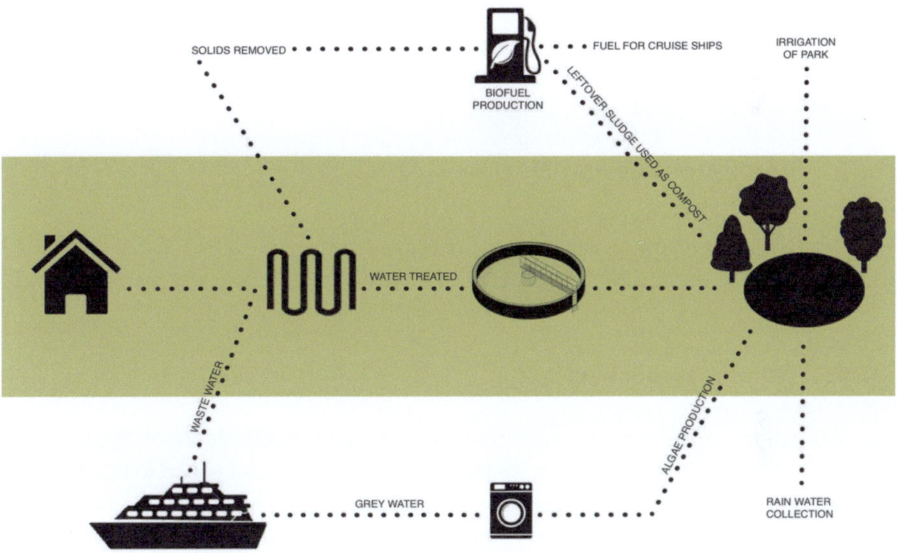

Fig. 9.14 Cruise ship waste water treatment scheme (City-zen Roadshow team 2016; image by Greg Keeffe, Queens University Belfast)

Fig. 9.15 Energy renovation by post-insulation and PV facades (left) or algae facades (right) (City-zen Roadshow team 2016)

On top of the ridge behind Gruž an investor planned a poorly supported plan for a golf resort. Golf courts are seldom sustainable – the one on Samsø is, by the way – but putting wind turbines between the greens improves the acceptability and productive capacity of Gruž.

A developed country is not a place where the poor have cars, it is where the rich use public transportation, said Petro Gustavo, mayor of Bogotá. Or bicycles, in cities such as Delft. In Dubrovnik cruise ship tourists come to the city centre by diesel bus, contributing further to fossil energy consumption and air pollution. A new tramway linking the Gruž port and historic centre of Dubrovnik is proposed, and furthermore, bicycle lanes and walkways for the more fit citizens and tourists.

After doing all these measures, almost all of Gruž' energy consumption could be reduced to its own renewable production. Instead of the 1100 hectares required initially, only 36 remained. With a few more measures exceeding the boundaries of Gruž and clean cruise ships, the place would be energy producing and much more liveable.

9.4.2 The Zero–Carbon Sydney Workshop

In May 2017, a workshop was organised at the University of Technology Sydney focusing on zero-carbon Sydney.

9.4.2.1 Number Crunching

The workshop started with lectures by specialists and stakeholders of Sydney. The real work started when population, space and energy numbers needed to be crunched to workable figures and when the current situation had to be mapped and charted. With some natural gas for hot water and (limited) heating most of the Sydney energy used is electricity that comes from coal-fired power plants, which have a maximum efficiency of 42% and of which the qualification 'low emission' is still worse than any other energy source.

The overall energy use of Sydney households was deducted, for their house and for transportation. If we want to get rid of fossil fuels, the use of natural gas and petrol or diesel will not be allowed anymore.

9.4.2.2 Domestic Energy

Heating and cooling a house electrically is most efficiently done by a heat pump, for which the electricity – in terms of primary energy the quantity will be less than the current technical solution – needs to be produced sustainably. Replacing the fossil car by an electric vehicle by estimation reduces the energy for transport by a factor of four. This reduction in fossil fuels leads to an increase in electricity demand, which can be solved predominantly by solar power. The average value of insolation on a horizontal plane in Sydney is very high, around 2 kW/m^2, leading to a potential production of 250 kWh per square metre of PV panel.

The minimum PV roof required to produce the electricity of the house and to compensate for the use of gas and petrol – which would make the household carbon and energy neutral – would amount up to 122 m^2, which is unrealistic and expensive. If the household would be all-electric, this area could be reduced to 57 m^2. With an energy-efficient home and better use of public transport and bicycles it could be reduced to a minimum of 35 m^2 of PV, which is very well attainable.

For the entire Sydney region, assuming a quarter of the total residential roof area is suited for PV, these areas could be more than energetically self-sufficient by solar power. Bio-fermentation of organic waste could produce biogas too, but the quantity is very limited.

The old Victorian houses of Sydney could have PV tiles, such as those developed by Tesla, and apartment blocks could have tropical roofs – hovering above the current roof, keeping it cool – with PV cells.

Unfortunately, housing constitutes only a twelfth of the total energy use of Sydney, so industries, amenities and transportation are much more significant. Regarding these, an earlier study of the solar potential (Arup 2011) identified suited roofs in the central business district. Apart from these, Sydney has many more places that could have solar panels installed, industrial warehouses most notably.

9.4.2.3 Biogas and Storage

The students looked into different infrastructures already present. The present gas grid could be important because it can be connected to the biomass potential. Gas-powered plants near the coast could be connected to waste treatment plants and become a biogas storage (Fig. 9.16). Biogas, artificial methane or hydrogen may be important in the near future as renewable source for times when solar and wind are insufficient to produce all energy required, or to solve fluctuations in the intermittency of sun and wind. The main arteries of gas pipes would then also be an underground part of this energy storage.

9.4.2.4 Wind from the Ridges and Sea

Where a flat plane runs into a steep slope wind will accelerate, and higher wind speeds lead to greater yields of electricity with wind turbines. In Sydney the prevailing wind comes from mainly two directions, the South West and from North East. Instigated by staff, the students studied the ridges in the Sydney metropolis and selected the ones suited for the high-performance wind turbines (Fig. 9.17).

In the city there is a wind potential as well, from small urban turbines on high-rise buildings, but the yield of these is limited.

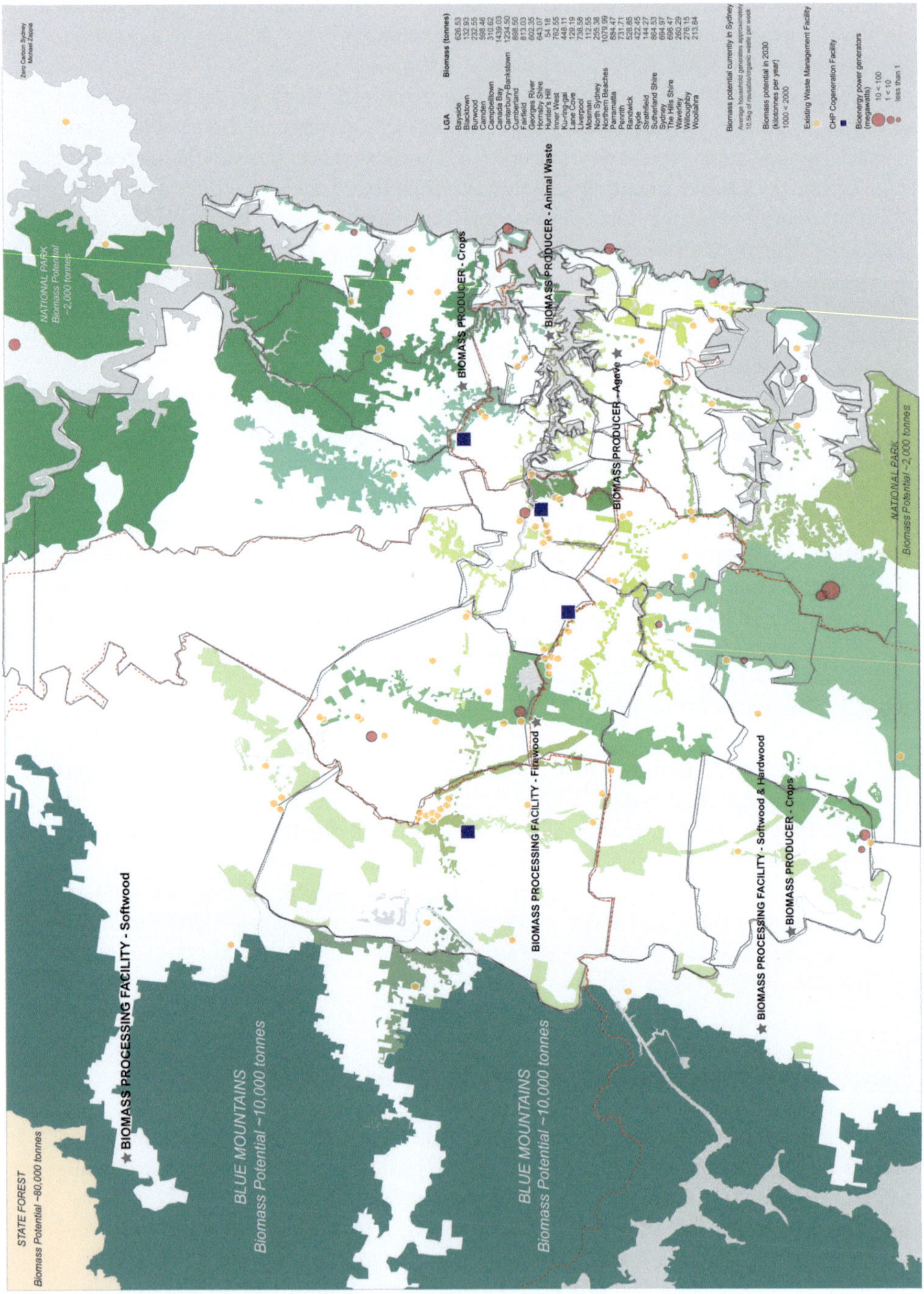
STATE FOREST
Biomass Potential ~80,000 tonnes
BIOMASS PROCESSING FACILITY - Softwood
BLUE MOUNTAINS
Biomass Potential ~10,000 tonnes
BLUE MOUNTAINS
Biomass Potential ~10,000 tonnes
BIOMASS PROCESSING FACILITY - Firewood
BIOMASS PROCESSING FACILITY - Softwood & Hardwood
BIOMASS PRODUCER - Crops
NATIONAL PARK
Biomass Potential ~2,000 tonnes
NATIONAL PARK
Biomass Potential ~2,000 tonnes
BIOMASS PRODUCER - Crops
BIOMASS PRODUCER - Animal Waste
BIOMASS PRODUCER - Agave
Zero Carbon Sydney
Michael Zappia
LGA Biomass (tonnes)
Bayside 626.53
Blacktown 132.93
Burwood 232.55
Camden 598.46
Campbelltown 310.62
Canada Bay 1439.03
Canterbury-Bankstown 1234.50
Cumberland 898.50
Fairfield 813.03
Georges River 602.35
Hornsby Shire 643.07
Hunter's Hill 54.18
Inner West 762.55
Ku-ring-gai 448.11
Lane Cove 129.19
Liverpool 738.58
Mosman 112.55
North Sydney 255.38
Northern Beaches 1079.99
Parramatta 884.47
Penrith 731.71
Randwick 528.85
Ryde 422.45
Strathfield 144.27
Sutherland Shire 864.53
Sydney 694.97
The Hills Shire 696.47
Waverley 260.29
Willoughby 276.15
Woollahra 213.84
Biomass potential currently in Sydney
Average household generates approximately 10.5kg of reusable/organic waste per week
Biomass potential in 2030 (kilotonnes per year)
1000 < 2000
Existing Waste Management Facility
CHP Cogeneration Facility
Bioenergy power generators (megawatts)
10 < 100
1 < 10
less than 1

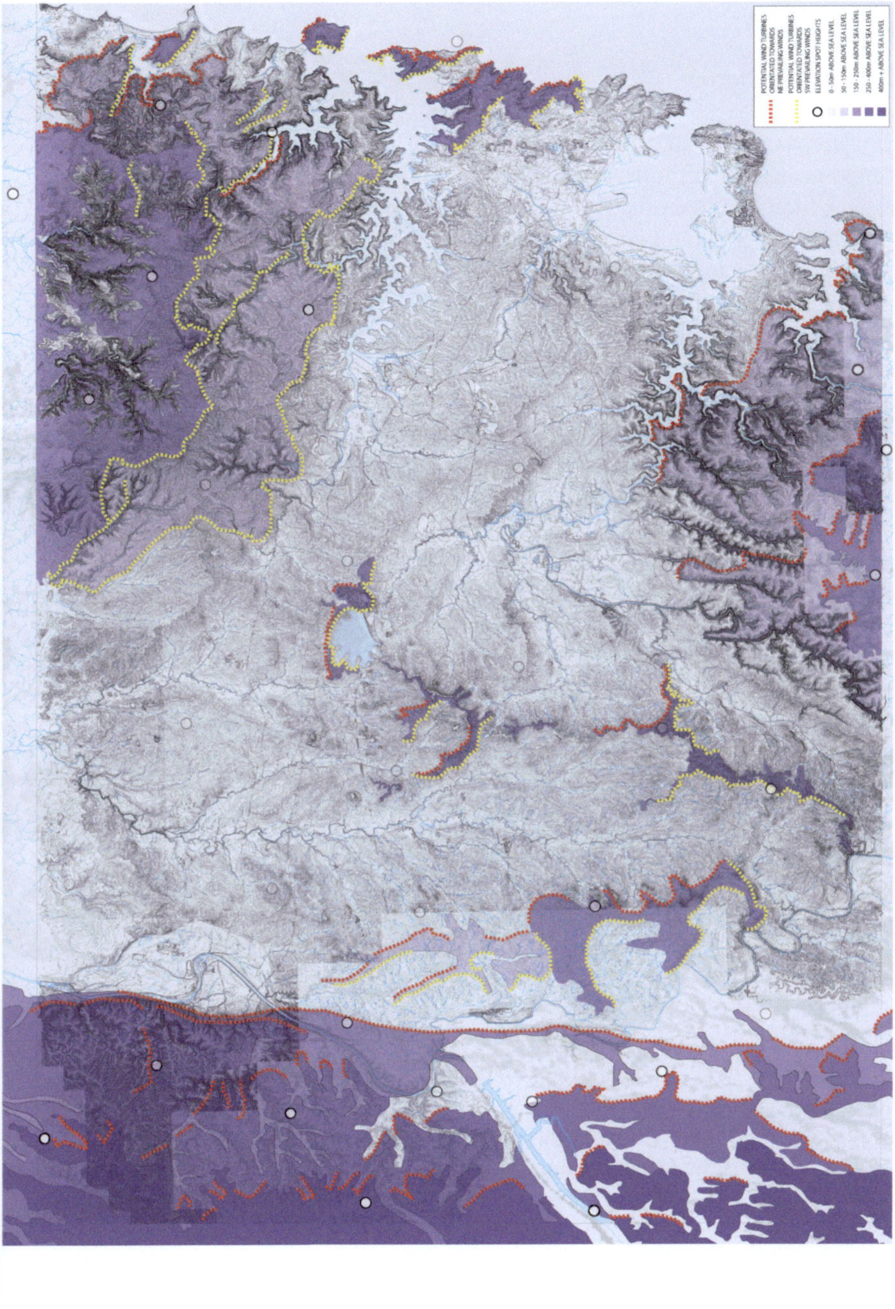

Fig. 9.17 Potential map of onshore wind energy from wind turbines placed on ridges in the Sydney region (Drawing: Hamish McKenzie, Zero Carbon Sydney team, led by Rob Roggema, UTS)

In contrast, more yield can be expected from offshore winds turbines, best placed not farther than 30 km from the coast, where the seabed makes a dramatic drop-off.

9.4.2.5 Water Power

Open water, salt or fresh can be used for heating and cooling, using heat pumps to exchange heat. Furthermore, there are possibilities to create power from reversed osmosis, which is called blue energy, at points where fresh water flows into the salt sea inlets. Another natural energy potential from the sea is wave power. This could be combined with the Sydney Barrier Reef (Roggema 2017), in order to protect Sydney's coastline and to create biodiversity around an artificial coral reef to compensate for the losses with the Great Barrier Reef near Cairns, 2400 km north of Sydney. Wave power vessels could be positioned in strategically positioned openings of the Sydney Barrier Reef, at places where conditions for surfing should be preserved or even served better. In action, wave power vessels would help to reduce the intensity of waves by tapping off energy. All energy potentials from water are depicted in Fig. 9.18.

9.4.2.6 Protection from Flooding

This protection from flooding and coastal erosion is rather urgent. In the week prior to the Zero Carbon Sydney workshop IPCC's prospects for sea level rise were raised to up to 2.5 metre by the end of the twenty-first century (Le Bars et al. 2017). Based on this value, the students drew dramatic maps depicting which parts of Sydney would be flooded, no storm surge needed. In order to protect Sydney from flooding the idea was conceived to create mangrove forests on the banks under threat of flooding, for the gradual deposition of sediments from the rivers between the mangrove roots. That way, with the sea level rising, the sediment would gradually build up a natural dike. It would lead to a green image of the Sydney sea banks (Fig. 9.19).

In a way, the proposed mangrove banks only give basic protection. In case of a storm surge, more drastic measures need to be taken. The Sydney sea inlet and harbour then need to be shut off from the sea. Therefore, the team designed a sea barrier with large operable shells – resembling the Sydney Opera House – that allow cruise ships and sailing boats to pass underneath (Fig. 9.20). This Sydney Ocean Barrier could be combined with a tidal plant so that two purposes are served in the meantime. It could become a proper touristic attraction while guarding Sydney's most important access to the Great Ocean.

Fig. 9.18 Potential map of energy from water in the Sydney region, also showing in grey the land that would be flooded with the latest prospected 2.5-metre sea level rise (drawing: Jelena Alavanja and Nadine Haddad, Zero Carbon Sydney team, led by Rob Roggema, UTS)

Fig. 9.19 Students' impression of the Sydney sea banks with flood-protective mangrove forest (Image: Jelena Alavanja, Zero Carbon Sydney team, led by Rob Roggema, UTS)

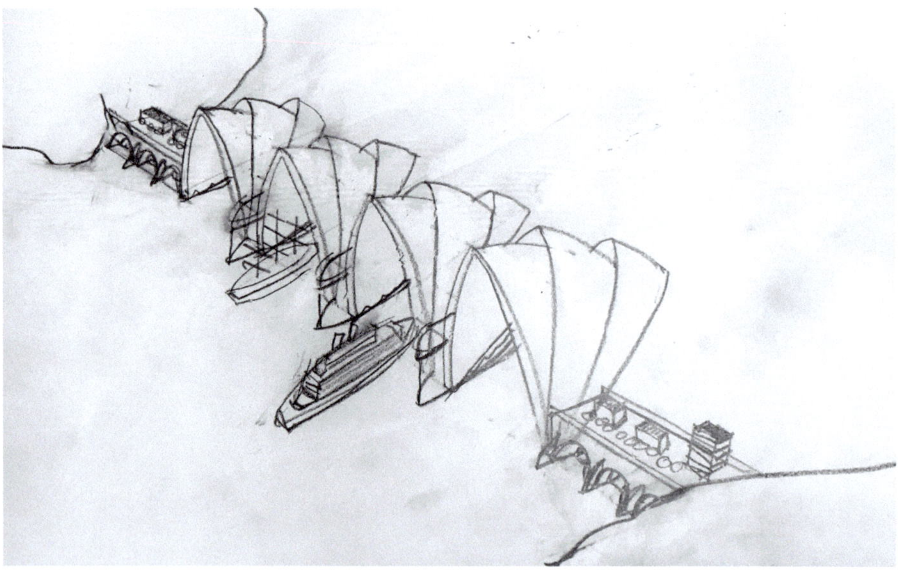

Fig. 9.20 Sketch of the Sydney Ocean Barrier (Zero Carbon Sydney team, led by Rob Roggema, UTS; drawing by the author)

9.5 Conclusion

9.5.1 Innovation, Integration, Imagination

There is a long journey of energy transition before becoming carbon neutral and it necessitates determined action, from the government, from the market, and not least from academics, planners and designers. A sustainable world needs a lot of innovation and innovation lies most in the no-man's land between disciplines. Moreover, the speed needed for timely adaptation requires collaborations between specialisms, combining academia with governance, industries and communities. And with the severe changes upon us, the world requires more imagination, because problems that are very different from the past have to be solved. Specialism would be paralysing.

9.5.2 The Importance of interactive Workshops

So, it is important that people understand more of different areas than their own and try to integrate them. The interactive workshops presented in this chapter – The City-zen Roadshow and Zero-Carbon Sydney workshop – are key therefor. Desk research, teleconferences, meetings or speeches will not help to find a solution like sitting together with a group of creative people with different qualities. Being forced to draw and calculate things enforces ideas to become practical and to visualise ideas as good solutions and not just words. Most importantly perhaps, they are fun to the people involved.

9.5.3 It Is about people and Communication

As Søren Hermansen stated, the problem is not technical, not even economical, it is really about people and their actions. The best example is from Samsø, where Sir Søren Hermansen encountered thresholds between people from a village, who did not dare to agree on the renewable energy plan until an important local person, the smith, had said that he thought it was a good idea. So social structures are very important, which are different everywhere, but convincing important key figures within a community can be decisive for success, positive or negative. Therefore, talking different languages, metaphorically, is the greatest asset of people who want to change the world. Somebody with a political agenda speaks a different tongue than someone from academia. Some people can be convinced by ratio, others by a beautiful fairy tale. Whichever of the two, a sustainable world is possible, and we need both qualities – hard data and imagination – to get there.

References

Arup (2011) City of Sydney renewable energy master plan – technical report. Arup, Sydney, Australia

Broersma S, Fremouw M, Dobbelsteen A (2013) Energy potential mapping – visualising energy characteristics for the exergetic optimisation of the built environment. Entropy 15(2):490–510

Graamans L, van den Dobbelsteen A (2017) Duurzaamheidsonderzoek Lidl Nederland (voorlopig rapport). TU Delft

IPCC (Intergovernmental Panel on Climate Change) (2014) Climate change 2014: impacts, adaptation, and Vulberability. IPCC, Switzerland

Keeffe G (2009) Synergetic city: urban algae production as a regenerative tool for a post-industrial city. In: Proceedings of IFHP 09, Berlin

Kleerekoper L, van den Dobbelsteen AAJF, Hordijk GJ, van Dorst MJ, Martin CL (2015) Climate adaptation strategies: achieving insight in microclimate effects of redevelopment options. Smart Sustain Built Environ 4(1):110–136

Klemm W, Heusinkveld BG, Lenzholzer S, Jacobs MH, van Hove B (2015) Psychological and physical impact of urban green spaces on outdoor thermal comfort during summertime in the Netherlands. Build Environ 83:120–128

Labo Ruimte, Posad, Universiteit Gent, Resource Design, 3E, Vlaamse Overheid (2016) Rapport Atelier Diepe Geothermie. BWMSTR, Brussels

Le Bars D, Drijfhout S, de Vries H (2017) A high-end sea level rise probabilistic projection including rapid Antarctic ice sheet mass loss. Environ Res Lett 12(4):044013

McDonough W, Braungart M (2002) Cradle to cradle – remaking the way we make things. NorthPoint Press, New York

Roggema R (2017) The Sydney barrier reef: Engineering a natural defence against future storms. 16 May 2017. The Conversation. http://theconversation.com/the-sydney-barrier-reef-engineering-a-natural-defence-against-future-storms-76862

Taleghani M, Sailor D, Tenpierik M, van den Dobbelsteen A (2014) Thermal assessment of heat mitigation strategies – the case of Portland State University, Oregon, USA. Build Environ 73:138–150

Tillie N, van den Dobbelsteen A, Doepel D, de Jager W, Joubert M, Mayenburg D (2009) Towards CO2 neutral urban planning – introducing the Rotterdam energy approach and planning (REAP). J Green Build 4(3):103–112

van den Dobbelsteen A (2008) Towards closed cycles – new strategy steps inspired by the cradle to cradle approach. In: Proceedings PLEA 2008 – 25th conference on passive and low energy architecture. UCD, Dublin

van den Dobbelsteen A (2017) Visiedocument City-zen Roadmap Amsterdam (voorlopig rapport). TU Delft

van den Dobbelsteen A, Lassen AK, Fremouw M, Goosen H, van Steekelenburg M, Peen M (2008) Het 1,3-meterplan – Hotspot Zuidplaspolder. TU Delft/Provincie Zuid-Holland, Den Haag, Netherlands

van den Dobbelsteen A, Keeffe G, Tillie N, Roggema R (2010) Cities as organisms: using biomimetic principles to become energetically self-supporting and climate-proof. In: Teng J (ed) Proceedings ICSU 2010 – First international conference on sustainable urbanization (CD-rom). The Hong Kong Polytechnic University

van den Dobbelsteen A, Jonathan T, Kruizinga J (2015) Prêt-à-Loger: zero-energy home with maximum living quality Increase. In: Proceedings of architecture and resilience on a local scale (AR2015). University of Sheffield

van den Dobbelsteen A, Martin CL, Keeffe G (2017) Zero-energy me – the struggle for individual energy Neutrality. In: Proceedings PLEA 2017 – design to thrive, Edinburgh

van der Hoeven F, Wandl A, Demir B, Dikmans S, Hagoort J, Moretto M, Sefkatli P, Snijder F, Songsri S, Stijger P, Yakovleva P, Wijtsma D, Blocken B (2014) Sensing Hotterdam: crowd sensing the Rotterdam urban heat island. In: Spool, vol 1. no. 2

WHO (2007) Quantifying environmental health impacts. Deaths and DALYs attributable to outdoor air pollution (2002). URL:http://www.who.int/quantifying_ehimpacts/national/countryprofile/mapoap/en/. Accessed 1 Dec 2017

Chapter 10
Transforming Transport, Transforming Urban Spaces

Jay Stricker

Abstract The coming together of new technologies, changing attitudes to car ownership and innovative business models provides the opportunity to transform personal and business transportation. Australia is one of the most urbanised countries in the world with historical urban form in major cities that has geographically separated residential areas from areas of employment. There is now an opportunity to reduce the currently high reliance on personal car ownership with new models of mobility based on the provision of mobility as a service incorporating all transport modes. A result of this would be changes to our urban forms that would create more living spaces for communities, replacing car parks and transforming suburban streets to social places.

10.1 Introduction

Australia is one of the most urbanised cities in the world with more than 80% of its population living in cities along its coastline. Sydney is the most congested city in Australia and New Zealand and is ranked the 29th most congested city globally (TomTom undated). Traffic congestion cost Australia AUD16.5billion in 2015, a 30% increase since 2010, and is forecast to reach AUD50 billion by 2031 (Australian Infrastructure Audit 2015). However, traffic congestion is a global phenomenon, impacting on the quality of life of urban dwellers. It is time to re-assess our concepts of mobility and the relationship between personal transport and urban form.

J. Stricker (✉)
AECOM, Sydney, NSW, Australia
e-mail: jay.stricker@aecom.com

R. Roggema (ed.), *Contemporary Urban Design Thinking*, Contemporary Urban Design Thinking, https://doi.org/10.1007/978-3-319-91950-8_10

10.2 Our Relationship with Private Cars: Myth and Reality

For the past three or four generations, owning a car has been synonymous with personal freedom. For most urban dwellers, this is a myth (Fig. 10.1). Private car ownership has been synonymous with the freedom to go anywhere, and to not be reliant on public transport timetables (ABS 2013; Lee 2003), with growth in motor touring in NSW as early as 1935 (Broomham 2001; NRMA 2017). Australia's urban form after World War II was focused on low-density sprawling suburbs, designed for private car access but leaving a legacy of car dependence which public transport is struggling to overcome. Between 1955 and 2013, ownership of passenger vehicles grew at a steady rate of 4% per annum ABS 2013).

The reality is urban drivers are generally trapped in congestion at least 6 days a week (Fig. 10.2). In fact congestion on the weekends in Sydney is similar to weekday peak hours (AECOM 2017).

Private car ownership is eroding the quality of life of in cities. Car ownership costs are high, with the average household in Australia spending around $AU332 per week on car ownership costs (NRMA 2017). Thirteen million private vehicles sit idle for 22 h per day in Australia. Large areas around transport hubs and commercial centres are dominated by car parks and streets are narrowed by on-street parking.

Driving on congested roads uses precious time that could be used for productive work or for leisure and family time. This contributes to reduced liveability, particularly where there is a large spatial separation between residential and employment

Fig. 10.1 The concept of freedom provided by car ownership

Fig. 10.2 The daily reality of car ownership in cities

areas. There is a significant spatial discrepancy between areas of population growth and employment growth in Australian state capital cities (Kelly and Donegan 2015). Between 2006 and 2011, more than half of Australia's population growth occurred in suburbs more than 20 kilometres from the city centres of the five biggest cities, Sydney, Melbourne, Perth, Brisbane and Adelaide. However, in the same period, more than half of the growth in jobs occurred less than 10 kilometres from the centre centres (Kelly and Donegan 2015).

10.3 The Mobility Challenge

Sydney's population has just reached 5 million, faster than predicted, and if this current trend continues, will reach 8 million before 2030 (Department of Planning and Environment NSW 2016). Like many cities around the world, Sydney is faced with the challenge of providing mobility and access to safe and sustainable transport on already congested roads for our growing population. Obviously public transport is part of the solution and there is currently an increased investment in public transport but it is not sufficient to provide for the variety of users and destinations required by the growing population.

We need both public and private transport to change, but we also need behavioural changes. Fortunately we don't need to wait for future solutions – the future is here -it's just not well distributed and understood. Both technology and mobility business models are being re-imagined and re-invented. The pace of change is as rapid as the transformation from 5th Avenue New York being dominated by the horse and buggy in 1901, to the dominance of the automobile in 1913.

10.4 Enabling Technologies

Technologies are being re-imagined and re-invented that enable rapid development of new vehicle and business models for mobility. The change in communication from landlines to smartphone and the connectedness provided by the Internet are two of the most important technological changes in recent decades. They are major enabling technologies, which are already changing the way we use transport. Global positioning systems (GPS) provide a platform for navigation systems, but it is the connectedness that smartphones and the internet tools provide that has enabled completely new sharing business models to develop, and in the case of transport, not only real time traffic and transport information, but the ability to book and access personalised transport.

10.5 A Subscription Service for All Transport Modes?

Imagine if all your travel, whether by private car paying road tolls, car-share, ride-share, bike-share, public transport and even parking, were all tracked by your smartphone and usage fees provided in one monthly or fortnightly account. This would provide greater transparency of your usage and enable you to refine your planning for how best to travel. This is now possible using smart devices and the Internet. Smartphones are used for rideshare and car-share bookings and access, and could replace the current magnetic cards used on public transport.

These technologies, along with improvements in spatial positioning, radar and camera technologies have also enabled biggest evolution in on-road transport; the development of connected and autonomous vehicles.

10.6 Accessible Autonomy

The age of accessible autonomy has arrived – and automated and connected vehicles, cars, buses and trucks - are on roads now. All new vehicles will have driverless capability within 10 years (Roads Australia 2017), however because of the varying take-up rates for new technologies, the time taken for fleet turnover and the time needed for mature regulatory frameworks to be developed, there is likely to be a period of around 20 years with a mixed fleet of driverless and human driven vehicles (Anderson 2017; Roads Australia 2017). The Australian and state governments are preparing policy and regulatory frameworks to enable automated vehicles to operate on Australian roads and trials of both cars and shuttle buses have been implemented successfully (Lyon et al. 2017). Autonomous vehicle trials are commonplace around the world (Anderson 2017). A dedicated radio frequency bandwidth has been set aside for vehicles and Geosciences Australia is refining the accuracy of locational data to support automated navigation.

10.7 Levels of Autonomy

Five levels of autonomy are generally recognised (ADVI undated) with the anticipated time periods for implementation (Venturebeat 2017):

- Level 1 Driver assistance –assistance provided for either steering, acceleration or braking
- Level 2 Partial – specific tasks automated with human driver monitoring – e.g. parking assist
- Level 3 Conditionally automated - system drives vehicles for sustained periods but a human driver must be alert for any system failures and intervene if needed, such as automated driving on specified sections of motorway and highways.
- Level 4 Highly automated – removes the need for a driver however the vehicle cannot fully operate under all conditions. The system drives in defined places or for sustained periods. No driver intervention is required. Examples are Valet parking and driverless shuttles that operate within a defined area.
- Level 5 Full automation – the system undertakes all driving tasks under all conditions with no steering wheel or pedals in the vehicle

The inflexion point for autonomous vehicles was around 2009 – autonomous mining trucks and trains commenced operation in the mines of the Pilbara region in Western Australia (Rio Tinto undated) and Google put a level-3 autonomous car on the road in California (Lyon et al. 2017). Other companies such as Tesla and Daimler were also developing automated vehicles (Fagella 2017). Fully automated vehicles are expected to be on roads around the world in 2025.

Looking back over the past 50 years, there has been a steady progression in automation of vehicles, firstly with automated gear change and cruise control and then with the development of on-board computers to automated parking and braking assistance. Currently, on-board computers are connected to GPS and to home bases by WiFi, enabling fully automated driving, but also using on-board cameras and sensors, including infrared and radar (Lyon et al. 2017).

The exact timing of fully connected, co-operative and automated vehicle roll-out will vary around the world but a common estimate is around 10 years for fully connected, co-operative vehicles to be functioning in fully autonomous mode but automated vehicle taxi and delivery fleets will become more prevalent from around 2021 as 5G digital networks are strengthened.

There are obvious benefits of automated vehicles:

- Road safety – less trauma on the roads. Automated vehicles have highly advanced sensors for objects and people around them and the on-board computers manage avoidance strategies. When the digital systems are mature they will be connected to each other so avoidance of crashes is relatively easy when the car knows exactly where other vehicles are on the road. This is a major benefit in terms of both the cost of road trauma and the impact of road trauma on individual lives and communities.
- Alternative uses of travel time – leisure or productivity. Without having to focus on the driving, passengers can choose between using their travel time for leisure using the on-board entertainment systems or to increase their productivity by working
- Social equity - Fully automated vehicles can provide equitable access to mobility for everyone including the elderly and disabled, providing significant social benefits.
- Potential environment benefits may be achieved as vehicles are becoming cleaner but there are also benefits if the total vehicle kilometres travelled is reduced. Increasing the use of electric vehicles will also help although that is only a benefit if there are sufficient charging stations and the power production is from clean energy sources. For example, the world's largest car-share company, Car2Go, established a fleet of electric car-share vehicles in California however the network of charging stations was not developed as rapidly as expected, so they were left with 40% of their fleet unavailable, at any time. They then converted most of their fleet to petrol vehicles. So there needs to be certainty that the appropriate infrastructure is provided on time.

10.8 Will Automated Vehicles Solve Our Congestion Problem?

It is not the technology that will solve our congestion problem, it is how government, societies and individuals respond to it and use the opportunities technology offers that can solve the congestion problem. Technology is the tool but behavioural change is needed to drive change.

Different models of mobility are needed. If car ownership continues as it is, the only benefit will be in road safety but not congestion, or reduction in car dominance of urban spaces (Fig. 10.3).

A move is necessary from car ownership to shared vehicles, and from active driving to autonomous. This will be a significant challenge for some people who identify

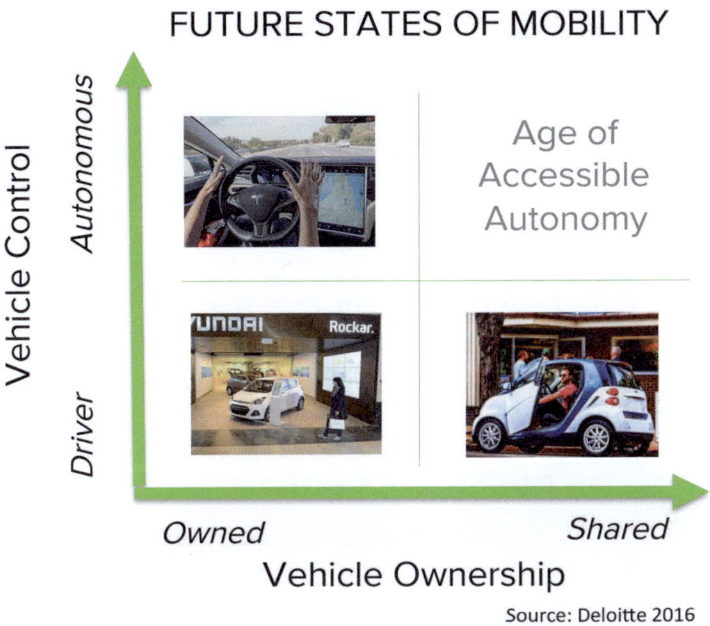

Fig. 10.3 The evolution of mobility. (Adapted from Corwin et al. Deloitte 2015)

with their vehicles – the old idea of certain types of vehicles as status symbols. There are also people who just really enjoy the activity of driving, although probably not so much on congested city streets. Australia is a big country, the tyranny of distance, combined with low population density, and with urban sprawl as the historical growth pattern for cities.

In Australia, as in the United States, transport choices have been dominated by car ownership for a long time – much more so than in Europe where population density has supported more rapid development of effective public transport and distance travelled are significantly less. For these reasons, government and industry will need to actively promote a change not only from private car ownership, but also to car-share and rideshare models.

10.9 Future Scenario's

Transport for New South Wales (NSW) has identified four scenarios in their 2016 Future Transport Technology Roadmap (Transport for NSW 2016).

10.9.1 Scenario One: My (Autonomous) Car Is Still King

In this scenario, automation drives a fundamental shift in door-to door transport services for people and goods, which induces demand, i.e. more private and commercial vehicles on the roads and vehicle kilometers travelled continues to rise. Congestion continues.

10.9.2 Scenario Two: We're All in This Together

The sharing economy has continued to gather strength and technology disrupts the model of individual vehicle ownership with a strong shift towards mobility on demand services. People use a mix of different transport options based on cost, convenience and their particular needs and preferences, using a mobility service account.

10.9.3 Scenario Three: Super-Commuting with Public, Active and Shared Transport

A wave of next-generation technology investments delivers dramatic improvements in public transport with sustained levels of superior customer experience. Dynamic timetabling is possible and is responsive to fluctuating demand caused by weather and events. This drives a significant increase in use of mass transit, but the road network remains important enabling the freight task and road based public and shared transport services.

10.9.4 Scenario Four: Why Travel So Much?

Personalised services delivered door-to-door have improved dramatically and are significantly more affordable, reducing the need for people to travel as much as previously. The spread of cloud computing and digital presence technologies along with community activity hubs, has given people more flexibility than ever to choose where they work, learn and relax. Most choose to do so as much as they can closer to home. This reduces the use of the road network for private travel and frees up some road space for other uses.

Obviously government needs to develop the appropriate policies to encourage multi-modal mobility to realise the benefits of the new technologies, but government can't, and should not, do that alone. It needs the active involvement of the community.

Fortunately there is a generational change happening in car ownership with less young people obtaining drivers licences, and choosing to use public transport and/or Uber for their journeys. In 2009, only 51% of people aged 20–24 held a driver's licence, compared with 70% in the nineties. In Sydney in particular this trend is driven by the high cost of real estate, so that car ownership is unaffordable for many young people.

10.10 Mobility as a Service – Bespoke Mobility

We need to rethink our relationship with cars to move beyond the private ownership and to think of transport as mode agnostic and select the most appropriate mode for each journey. Business models that provide mobility as a service would be supported by a subscription based service, using rideshare and car-share as well as public transport – a multi modal mobility model. The transformation of transport systems to user-driven and contestable consumer services has already begun with the rapid expansion of car-share and rideshare services. Lyft is extending it's rideshare services to an additional 100 US cities in 2017, increasing its target customer base to 72% or 231 million people (Anderson 2017). The recognition of private vehicle ownership as an expensive and inefficient transport mode is increasing globally, and particularly, in congested cities. Lyft predicts that the majority of US citizens living in major metropolitan areas will no longer own a car by 2025 as LyfT, Uber, Car2Go and similar services dominate the mobility market (Anderson 2017).

The trend to increased use of car-share and rideshare is forcing public transport authorities to develop more responsive transport systems than the current emphasis on mass transit. Both the United States and United Kingdom report reductions of 8–9% in mass transit patronage (Roads Australia 2017) due to the attraction of affordable on-demand services and ridesharing.

We need to make public transport more responsive and better integrated with other travel modes. An example of responsive public transport is the Denver Call'n Ride system (GoDenverapp.com) - a personalised bus service that travels within selected areas. These 14 seater buses are on call by a smart device or you can wait at designated stops for the frequent services. There have been similar services operating in some Brazilian cities such as Porto Allegre, for more than a decade, and they have been very effective in providing the first and last mile between home and major transport hubs or commercial or medical centres (Author's study tour November 2003). In some cities such as Sao Paulo, there are specialised services used for people with special needs, and their carers, and for people who are ill or frail. These services have drivers, and sometimes a second person, trained in first aid to assist passengers with medical conditions (USICD 2013). A US State Government department contracts Uber to provide cost-effective mobility services for their disabled customers and hospitals in the US and Australia are starting to use Uber services for transporting outpatients to hospital for treatment (Roads Australia 2017).

Rideshare services can be tailored to be complementary to mass transit systems by providing first mile/last mile transport between home and a mass transit hub. If this were to be used in Melbourne or Sydney, the large commuter car parks at outer suburban train stations, where cars sit unused all day while their owners commute to the city to work, could be used for more productive land uses. Carshare, rideshare and public transport are complementary in reducing usage of private vehicle but need appropriate government policies and collaboration between government and industry to be truly effective.

AECOM undertook research into opportunities for more sustainable transport and the potential benefits for our cities (Jackson et al. 2016). In partnership with GoGet, the largest car-share provider in Australia, and the University of New South Wales, AECOM analysed GoGet vehicle usage data to forecast car share growth and its impact on peak car, which is the point in time at which private car ownership will stop rising and start declining.

Data from GoGet shows that:

1. Transport on demand or provision of mobility as a service, will reduce vehicle kilometres travelled on our city roads
2. 88% of GoGet members no longer own a car
3. GoGet members travel around 2000 kilometres less per year than car owners
4. About 20% of privately owned car driving is searching for a parking space

The international literature review by Jackson et al. revealed the typical characteristics of car-share users (see Fig. 10.4) and these were benchmarked against empirical data on car-share use in Australia, supplied by GoGet.

A predictive model was developed to estimate the future potential for car-share.

- The model uses census data and projections to map the census zones, which meet those criteria.
- By estimating how many people live in these zones, it was possible to estimate the potential market for carshare users and through some assumptions, how many of these would join a carshare scheme. The assumptions are based on membership and car usage data from Australia's largest carshare company which indicates that the most likely group to switch from car ownership to carshare are those living in high density inner metropolitan areas
- The number of privately owned cars in Sydney was forecast, using data from the NSW Bureau of Transport Statistics and NSW Roads and Maritime Services, to provide a benchmark.
- Research by Phillip Boyle & Associates (2016) was used to estimate benefits, including;
  - Number of private cars each car-share vehicle removes from the road (10);
  - Space required to park each private car (13.5 square metres) and
  - Reduction kilometres driven each year by car-share users compared to private owners (estimated number of car-share users times 2000).

The results are shown in Fig. 10.5.

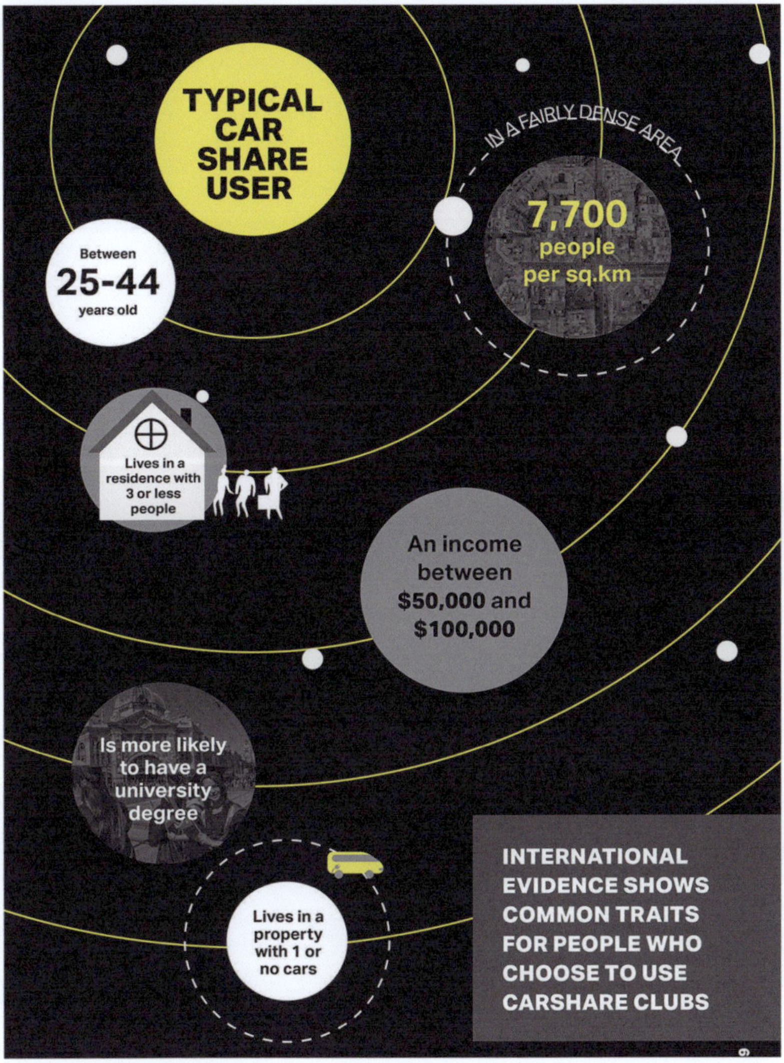

Fig. 10.4 Characteristics of car-share users. (Source: Jackson et al. 2016)

If the potential expansion of car-share in Sydney is achieved in 2036, there would be 91,000 fewer cars on the road, up to 1.2million m^2 of more urban space and over 180,000,000 less vehicle kilometres travelled.

To give this some context, 91,000 equates to 2% of all cars in Sydney by 2036 – which may seem small, but there is a noticeable reduction in congestion and travel times during school holidays in Sydney and that is around a 5% reduction in cars on the road (NRMA Motoring & Services 2013). With the right policy levers in place, and community education on reducing the burden of car ownership, we may be able to expand car-share usage beyond our projections.

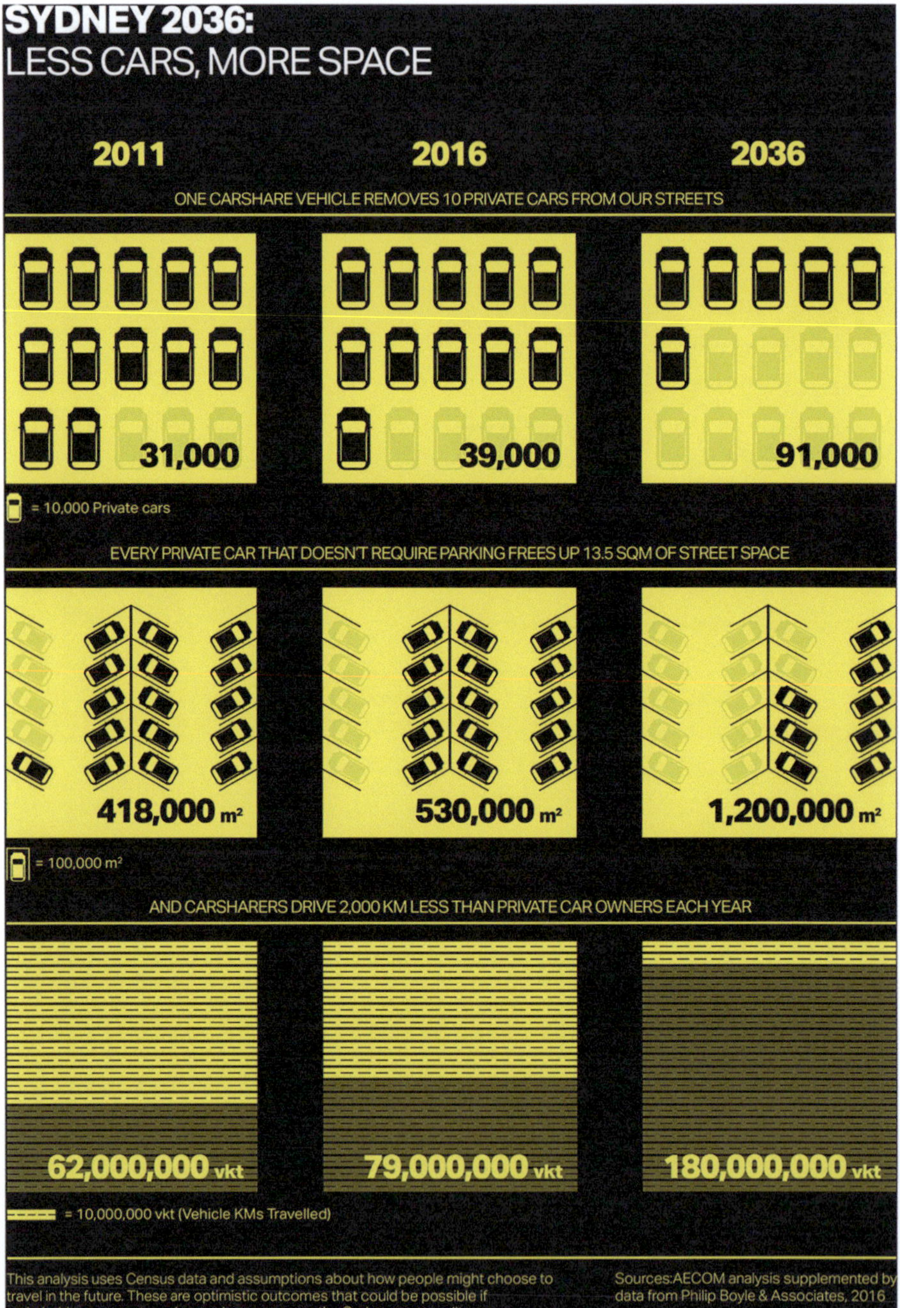

Fig. 10.5 Key findings from the modelled potential of carshare use in Sydney. (Source: Boyle and Associates 2016)

Business models are also evolving. The rise of car-share, again enabled by smart technology, is evolving further to one-way trips – park anywhere - free-floating cars. These can be used point-to-point for one-way trips rather than needing to return the cars to their dedicated car space. Point to point car share, and bike share, are well established in Europe and North America (Roads Australia 2017).

These mobility business models could transform the way we move around Australian cities, just like they are starting to make an impact in North America and Europe. The big game changer is not automation of the driving task, although that will provide social and equity benefits, but the change from private car ownership to rideshare.

10.11 What Does This Mean for Our Road Networks and Urban Spaces?

Large commuter car parks at rail and bus interchanges, airports and shopping centres will not be needed anymore. These could be reconfigured for other, more beneficial land uses. Residential buildings won't need car parks in their basements, but will need well designed, safe pick-up/drop-off points for our automated car or bus services.

These changes are already happening with some future-oriented developers constructing apartment buildings either with no or very few private car spaces. GoGet has been integrating car-share into residential developments since 2009 and offers discounted car-share membership for residents. The city of San Francisco is considering car-parking reforms, including making some new buildings ineligible for parking permits. The City of Sydney, a leader in parking reforms and support for cycling networks, introduced this reform in 2009.

The new generation of motorways under construction in Sydney and Melbourne are being designed to support automated and connected vehicles. Even if a reduction in vehicle movements on our roads is achieved, we will still need the high quality motorway and arterial networks to separate the long distance traffic and the freight vehicles from the local streets. Local streets may become narrower without the need to provide for on-street parking, but with identified pick-up and drop/off bays for whichever on-demand transport mode we require; cars or buses. The additional street space can be used to increase the functional value of streets as social spaces incorporating market places and community gardens.

Currently there are around 6500 fuel (service) stations across Australia – mostly in strategically significant locations (Lang and Rosenwax 2017). In December 2016, BP paid Woolworths almost AUD 1.8 million for 527 sites, indicating that they are very valuable properties. Their real future value is not the fuel they provide, although they will continue to offer that in some form for at least the next 20–30 years – it is their locations. Generally these sites are large, well connected to the road network and positioned in an urban location close to businesses and with good access from residential areas. As vehicles are becoming more fuel-efficient and changing fuel from hydrocarbon sources to electric, these service stations need to evolve to match the market. Many now are incorporating better quality food outlets than have traditionally been provided, as well as retail stores, effectively becoming a one-stop shop for customers. Alternative uses, in addition to providing convenience offerings and charging points for electric vehicles, could be as hyper-connected shared workspaces or maker spaces of the future, increasing the opportunities for local employment and diversity.

10.12 Conclusion

The building blocks for future more liveable cities and sustainable transport are in our hands. To stack them appropriately to achieve the benefits of the ride-sharing revolution will require collaborative effort from thought leaders, governments and mobility businesses, and a public that is sufficiently well informed and supported to accept the new world of mobility as a service, automated or not.

References

ABS (2013) Car nation report 4102.0 Australian social trends Australian government. http://www.abs.gov.au/AUSSTATS/abs%40.nsf/Lookup/4102.0Main%2BFeatures40July%2B2013. Accessed 19 Oct 2017

ADVI (undated) Levels of driverless cars. http://advi.org.au/wp-content/uploads/2017/06/Levels-of-driverless-car.pdf. Accessed 3 Oct 2017

AECOM (2017) WestConnex M4-M5 link environmental impact statement roads and maritime services NSW, appendix H: technical working paper: traffic and transport p z43

Anderson, D. (2017) Shared, connected and autonomous vehicles report on the roads Australia study tour 29 march- 13 April 2017, National Transport Commission (NTC) 2017

Australian Infrastructure Audit (2015) Our infrastructure challenges, executive summary, April 2015

Boyle P, Associates (2016) The impact of carshare services in Australia, International Carsharing Association.

Broomham R (2001) Vital connections – A history of NSW roads from 1788. Hale & Iremonger in association with the Roads and Traffic Authority NSW, Alexandria NSW.

Corwin S, Vitale J, Kelly E, Cathles E (2015) The future of mobility. Deloitte University Press https://dupress.deloitte.com/content/dam/dup-us-en/articles/transportation-technology/DUP-1374_Future-of-mobility_vFINAL_4.15.16.pdf. Accessed 27 Oct 2017

Department of Planning and Enviroment NSW (2016) 2016 NSW population and household projections. http://www.planning.nsw.gov.au/Research-and-Demography/Demography/Population-projections. Accessed: 3 Oct 2017

Fagella D (2017) Self-driving car timeline for 11 top automakers. VentureBeat. https://venturebeat.com/2017/06/04/self-driving-car-timeline-for-11-top-automakers/. Accessed 30 Nov 2017.

GoDenverapp.com. Accessed 08/04/2017

Jackson R, Wu X, Rosenwax J, (2016) Transport on demand – Accelerating Australian cities, AECOM Australia 2016

Kelly J-F, Donegan P (2015) City limits: why Australia's are broken and how we can fix them. Melbourne University Press March 2015

Lang A, Rosenwax J (2017) Fuelling the future. http://www.aecom.com/au/brilliantcityinsights/future-petrol-stations/

Lee R (2003) Linking a Nation: Australia's Transport and Communications 1788–1970. Australian Heritage Council. http://www.environment.gov.au/heritage/ahc/publications/linking-a-nation/chapter8. Accessed 10 Oct 2017

Lyon B, Hudson N, Twycross M, Finn D, Porter S, Maklary Z (2017) Automated vehicles do we know which road to take? Infrastructure Partnerships Australia.

NRMA (2017) The future of car ownership. National Roads and Motorists' Association

NRMA Motoring & Services (2013) NSW congestion busting strategy – Call for action

Rio Tinto (undated) Mine of the Future™. http://www.riotinto.com/australia/pilbara/mine-of-the-future-9603.aspx. Accessed 18 Sept 2017

Roads Australia (2017) Preparing for the driverless revolution transport reform policy chapter, Roads Australia

TomTom (undated) TomTom traffic index, measuring congestion worldwide. https://www.tomtom.com/en_gb/trafficindex/list?citySize=LARGE&continent=ALL&country=ALL. Accessed 3 Oct 2017

Transport for New South Wales (2016) Future transport technology, roadmap 2016. https://future.transport.nsw.gov.au/wp-content/uploads/2016/11/Future-Transport-Technology-Roadmap_2016__.pdf. Accessed 30 Nov 2017

USICD (2013) Improving accessibility to transport for People with Limited Mobility (PLM), A practical guidance note. http://www.usicd.org/doc/Accessibility0Report0Final.pdf. Accessed 30 Nov 2017

Venturebeat (2017) Self-driving car timeline for 11 top automakers, https://venturebeat.com/2017/06/04/self-driving-car-timeline-for-11-top-automakers/. Accessed 30 Nov 2017

Chapter 11
Australian Urbanism: Way Forward

Rob Roggema

Abstract In this book a broad range of aspects regarding Australian urbanism has been highlighted. Current practice, suggestions for improvements, necessities and views for the future have all been mentioned in the different chapters in this book. All authors don't take current practice for granted. They all touch on gaps, weaknesses and longstanding routines that stand in the way of becoming better Australian cities. There are several 'immediate' improvements that can be made. These range from habits and governance systems, thinking at a more conceptual level and integrating sustainability in planning.

11.1 Introduction

In this book a broad range of aspects regarding Australian urbanism has been highlighted. Current practice, suggestions for improvements, necessities and views for the future have all been mentioned in the different chapters in this book. All authors don't take current practice for granted. They all touch on gaps, weaknesses and longstanding routines that stand in the way of becoming better Australian cities. There are several 'immediate' improvements that can be made. These range from habits and governance systems, thinking at a more conceptual level and integrating sustainability in planning.

11.2 Land Ownership

The current urban development is caught in a trap of land ownership. Most of the land for development is owned by a whole range of different owners, from large sophisticated developers to individual residents. Where some of the larger

R. Roggema (✉)
Faculty of Design Architecture Building, University of Technology Sydney, Broadway, NSW, Australia

Cittaideale, Wageningen, The Netherlands
e-mail: rob@cittaideale.eu

R. Roggema (ed.), *Contemporary Urban Design Thinking*, Contemporary Urban Design Thinking, https://doi.org/10.1007/978-3-319-91950-8_11

developers understand the concept of value capture as result of increasing quality and sustainability of a development, the majority of both larger developers and individual land owners have a mindset in which return on investment implies earning money on the short term. In practice this has almost lead to an expectation, and entitlement to make a reasonable amount of money out of developing the land. This expectation forms a constraint for government to put strict rules in place around the type of development, the program to be realised, its quality and the level of sustainability. In each of these factors a risk may be hidden for the landowners of not receiving a profit on their investment in the land. Therefore it is extremely difficult for planners and designers to get grips on the desired future urban quality of an area. This dominating role of landownership should be brought back to reasonable proportions if the government is serious about the future quality of urban areas and neighbourhoods. However, this may require a fundamental transformation in regulatory system in Australia, and countries with similar planning systems.

11.3 Conceptual Design

Many of the current thinking in urban planning, especially at the scale of metropolises and larger urban concepts are dominated by a 'post-active approach'. Even the plan for metropolitan Sydney (Greater Sydney Commission 2017), which needs to be applauded because it is one of the first conceptual plans for this scale in Australia, even this plan is following developments that were happening already. The ongoing sprawl in Western Sydney is more or less sustained and made into spatial policy in hindsight. This turns the three city concept as if it is made a conceptual plan after the developments occur, instead of conceptual planning taking place before any new developments are realised. This way a large-scale conceptual design for the region could have directed urban development. A large-scale long-term vision is useless if it only follows existing urbanisation. For the third city, the area between the current urban boundary and Nepean river in the west, it means that sprawl is supported and around one million people are to be located in an area that is on average, in summer, too hot to live in. At the same time housing here is relatively cheaper than in the Eastern Harbour city, but energy bills will be rising as result of rising energy prices, but also due to increased use of energy to keep the houses cool in summer, and to drive the car(s) to the urban centres with jobs, and these are generally far away. From different viewpoint this makes living in this area relatively unsustainable, as the lifestyle of the majority of the residents will be above average energy usage and the outdoor environment will prove unhealthy. This is (partly) caused by the absence of a regional spatial concept before developments take place stealthily. And, do we really want to put people in the position to live in the sticky, unhealthy environment that this landscape tends to be when urban heat island effects are impacting the microclimate in the area?

Instead, if the new airport in Badgerys Creek happens to be really necessary, could it just be created amidst a green parkland, that mainly operates as 'landscape', an ecologically diverse site where native vegetation would be rehabilitated, and leisure and water storage would be the main function. This 'Green Heart' resembles

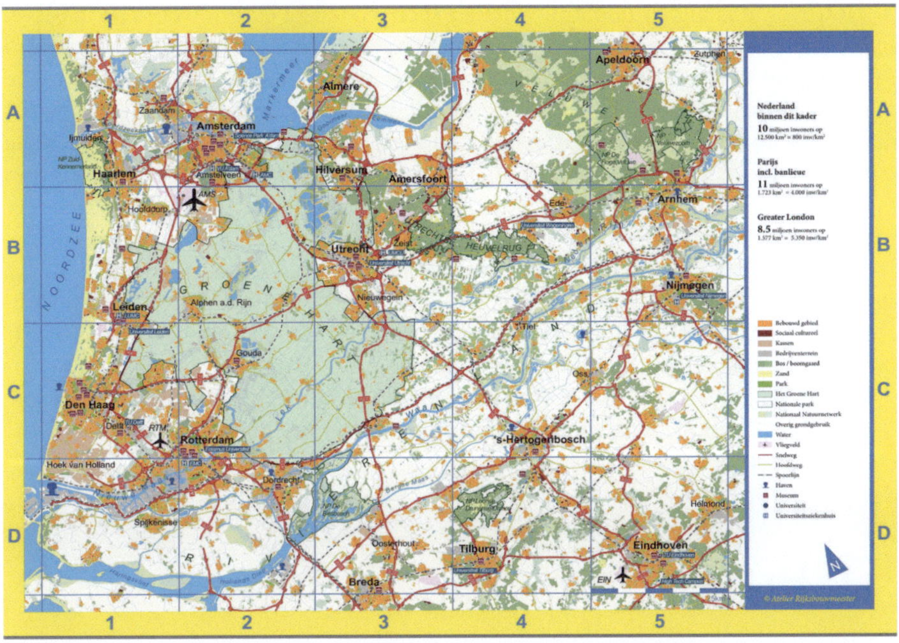

Fig. 11.1 Randstad in the Netherlands. (Source: Atelier Rijksbouwmeester)

Fig. 11.2 Central Park, Manhattan. (Source: mybestplace.com; http–//mybestplace.com/aroundtheworld/wp-content/uploads/2016/06/Central-Park-Manhattan-New-York-IMG-SLIDER-1.jpg)

the spatial concept that emerged over time for the Randstad Holland, where the large cities of Amsterdam, The Hague, Rotterdam and Utrecht surround a large open space, which is in use as nature, agriculture and leisure reserve (Fig. 11.1). Also, it reflects thinking that created Central Park in New York, an open parkland space amidst the high-rise buildings of Manhattan (Fig. 11.2).

However, it is highly doubtful whether such a concept design would be feasible, as the landownership in the area is such that the 'need' and 'right' of land owners to capitalise on their investments stands in the way of designing with a free mind. The main question is why? An alternative concept could provide the regenerative power that nature requires to survive the hot summers, the sometimes huge floods in the area and the occasional bushfire. It also could prevent noise and air pollution from the new airport from influencing new inhabitants in the area, set aside the safety levels it could enhance.

A timely and conceptual discussion about the spatial future of larger urban regions, such as Sydney metropolitan, could have led to better quality, and a healthier, more comfortable living environment for hundred of thousands of people. If only the thinking could be brought beyond the council boundaries and landownership, and put the features that make the region so beautiful, spectacular and liveable to the fore. To speak with the campaigner for the Bill Clinton campaign in 1992 (James Carville): it's the coast stupid!

11.4 Fix Current Problems

In the wake of a pro-active conceptual design process for the Sydney Metropolitan region, urban development should not be allowed to unlimitedly grow in direction we later might regret. In the meantime there is enough to 'fix'. Before even thinking about a third city the Parramatta River City (second city) should be improved and developed into a high quality urban environment so competitive people from the northern beaches or the inner west would consider living there. The accessibility of more and larger green spaces, the connectivity of this area with other suburbs and the Sydney Centre, and the quality of public spaces, such as the redesign of Parramatta square illustrates (Fig. 11.3) could all alleviate the current status of large parts of the second city as literally 'second best'. Increasing the density in this area could offer easily the capacity that is planned for the third city and 'done well' it contributes to urbanity of Parramatta and beyond. Only a high quality public space, well designed and with abundant liveliness could justify high-rise and high-density urban quarters.

The second 'fix' that can happen no matter a third city is transportation. Public transport needs to be frequent and fast and need to connect the region as a network, so passengers could move in the network from and to every neighbourhood direct and in a short time. Metro lines, light rail and supported by heavy rail, these lines would need to serve large amounts of people, connect the busiest hubs, and be prioritised over roads and slow trains. If this network is dense and lateral the efficiency will improve and usage of the public transport will increase. Additional measures as the accessibility of stations, the spatial quality of stations and the amount of bicycle parking in or extremely close to stations will further improve the use of public transport.

Fig. 11.3 Artist impression of the new Parramatta square. (Source: buildsydney.com; https://www.buildsydney.com/wp-content/uploads/2017/02/parramatta-square-render.jpg)

Thinking about the third city as an eventual future, public transport needs to be available before the development of this area has started. As we are already too late to realise this, the effort to realise fast trains and metros to and from the third city are conditional and should be realised and in operation in anticipation of eventual further urban developments, or the realisation of a new airport.

11.5 Governance

Governance is defined as the 'process of decision-making and the process by which decisions are implemented (or not implemented). Good governance has got eight characteristics: It is participatory, consensus oriented, accountable, transparent, responsive, effective and efficient, equitable and inclusive and follows the rule of law. It assures that corruption is minimized, the views of minorities are taken into account and that the voices of the most vulnerable in society are heard in decision-making. It is also responsive to the present and future needs of society' (UNESCAP undated). Many of these characteristics are not met in the Australian practice of urban development. For good governance different layers of government should work jointly together to reach the best outcomes for all their citizens, wealthy or not, more or less educated.

To perform good governance in practice the Sydney region has to align and establish collaboration between 30-odd councils, the State government of New South Wales and the Federal government, in the meantime dealing with vested interests, established land positions, a range of advocacy groups and the media. The forming of a regional thinker, the Greater Sydney Commission is a good step forward, but seems toothless as it comes to enforcing the policies to realise the goals set. For the trustworthy of a government it is important to make a vision, and keep working along the lines of that vision. With ever swinging governments, every 4 year the vision is redefined radically, which makes the reliability not extremely large. Therefore a 'governance revolution is required' (Williams 2018): 'The problem is structural and constitutional: essentially, state government is too powerful at a metropolitan level and local government is not powerful enough. Silos and poorly integrated tiers of government do not a great city make'. As Williams propagates, the Greater Sydney Commission should therefore be strengthened and must be made accountable for all Sydneysiders living in the entire Sydney Metropolitan region. This way the region could introduce a certain way of regional self-government, for instance by introducing a pan-Sydney forum, in which all mayors and State leaders congregate and make decisions for a sustainable and resilient, prosperous region. The Greater Sydney Commission would fall naturally under this new governance structure.

11.6 Landscape First

To survive as a species on earth, humans are dependent on natural features, from the plants and animals that provide us with new science to cure illnesses (Genome 10K 2009), the fertility and regenerative capacity of our soils to provide us with food, the availability of clean water or the capability to generate and provide energy, to name a few. Landscape plays an essential role in these processes, as it provides the space, the system and the connections to create resilient socio-ecological systems (Gu and Subramanian 2014). For urbanisation to start with the landscape is essential if one wants to increase resilience and sustainability in the city. The establishment of ecological relationships and the spatial connections in the landscape will fulfil the needs for providing the services to urban residents as mentioned before. Looking at the Third City for instance, the creek systems, its side creeks and gullies form the basic framework of these interrelations, and can provide the space not only to store water, clean water, produce food, and develop biodiversity, but give also the opportunity to residents to walk, cycle and stay in these spaces, or enjoy the view. Once these systems are taken as the starting point for design, and the flood prone areas are generously (ie. Beyond the 1:100 year flood boundary) incorporated in the framework, the city may fit in with this basic condition for growth. Neighbourhoods and transport facilities can be implemented while making sure the functioning of the landscape is uninterrupted. This way, urban living is paired with an ecological development, water management and a general high level of sustainability. Better places are created, which are healthier and more comfortable for people. Finally, when future change is already incorporated in the design for the region, climate impacts will harm less, because resilience is increased hence vulnerability decreased.

11.7 Design Together

In designing attractive cities it has long been the case to rely on professional expertise only. This has led to waves of urbanism to the 'liking' of generations of urbanists and architects. This process of the 'planner who knows' should be left behind (Gunder 2011). In order to make plans that are more appreciated by the users, we need to start making use of the expertise of local citizens. These people, often residing for a long period in a certain area, have a whole different perspective on the area, and can feed the design process in a different way. This way knowledge that has accumulated for many years in the area can become of use to the design process. This brings the design beyond the technical aspects of water systems, historic landscape structures, soil and geomorphology, calculations of traffic, and the required program for the area to issues such as the emotional bond people have with the area, the feeling they have developed over the years and the insights they have obtained by being in the area on a daily basis. The combination of both types of knowledge leads to collaborative design, co-design or co-creation of e new vision on the future (Fig. 11.4; Rli 2016).

The benefits are evident: people feel respected and taken seriously, even problematic issues can be discussed and brought to a solution, and a shared and supported vision for the future can be developed.

11.8 Conclusion

The current state of urbanism in Australia is in its infancy. Despite the fact that there is a tradition of thinking at the spatial higher scales, as is illustrated in chapter two, the current mainstream planning and design practice is one of following the established approaches, not challenging the constraints such as for instance instigated by landownership, the political short term orientation and has therefore a certain superfluity. Serious barriers to raise the quality of spatial design and the spatial quality realised in the streets, and squares of the major Australian cities need to be overcome. These comprehend, but are not limited to allowing for a different role of land ownership, a governance structure that needs to be overturned, the need for fundamental attention for the landscape as basic feature for urban development, the introduction of conceptual design and meanwhile fixing existing problems, and allow everyone to co-design their own urban environment.

Many people of which some have authored a chapter in this book, have great ideas and understand the necessity to adapt current urban development practice to more sophisticated approaches. Much can be learnt from planning processes experienced in countries with a long planning tradition. The challenge for Australian urban planning and design is to leapfrog these long-standing traditions, and anticipate future change in a way that it prepares society in a spatial way for the longer term future. The influence of data on the spatial quality of urban spaces and on the

Fig. 11.4 Innovative design process connecting different types of expertise. (Source: Rli 2016)

potential well-being of urban residents for instance is a field under construction, worth leapfrogging. But also the anticipation of uncertainties in climate impacts is an issue that requires anticipation in the planning and design profession, which, if done well, could lead to more advanced solutions than the longstanding planning cultures. This however requires courage, the courage to treat not only the planning and design profession in a different way, but also the courage to fundamentally change the relationship between government, development and landownership, and the way new governance structures are implemented.

Finally, the power of conceptual thinking requires a boost within the Australian culture of respect. Conceptual thinking enlarges the key focal points of a proposition, almost without nuance. Propositions that solve multiple problems with one concept, integrating problems in one design, but also neglecting details that for many are essential. The stubbornness of these propositions often feel as if they have a lack of respect in the minds of many Australians. But it is this respectfulness, which often looks like an exaggerated politeness, results in a discussion full of understanding each other points of view, but do not allow for a real debate, and this is exactly the reason why nothing ever seems to change. This 'deep culture' is difficult to change, but is worth tackling for the good of the liveability, resilience and sustainability of future Australian cities.

References

Genome 10K (community of scientists) (2009) Genome 10K: A proposal to obtain whole-genome sequence for 10 000 vertebrate species. J Hered 100(6):659–674. https://doi.org/10.1093/jhered/esp086

Greater Sydney Commission (2017) Our greater Sydney 2056 – A metropolis of three cities – Connecting people. State of New South Wales, Sydney. https://www.greater.sydney/draft-greater-sydney-region-plan. Accessed 14 Mar 2018

Gu H, Subramanian SM (2014) Drivers of change in socio-ecological production landscapes: Implications for better management. Eco Soc 19(1):41. https://doi.org/10.5751/ES-06283-190141

Gunder M (2011) Fake it until you make it, and then…. Plan Theory 10(3):201–212

Rli (2016) The connecting landscape. Council for the Environment and Infrastructure, The Hague http://en.rli.nl/sites/default/files/advisory_report_the_connecting_landscape.pdf. Accessed 16 Mar 2018

UNESCAP (undated) What is good governance. United Nations Economic and Social Commission for Asia and the Pacific, Bangkok http://www.unescap.org/sites/default/files/good-governance.pdf. Accessed 16 Mar 2018

Williams T (2018) Our cities need a governance revolution. The Fifth Estate. Published 13 March 2018. https://www.thefifthestate.com.au/columns/spinifex/our-cities-need-a-governance-revolution/98011. Accessed 15 Mar 2018

Index

R. Roggema (ed.), *Contemporary Urban Design Thinking*, Contemporary Urban Design Thinking, https://doi.org/10.1007/978-3-319-91950-8

S

T

U

V

W

MIX
Papier aus verantwortungsvollen Quellen
Paper from responsible sources
FSC® C105338

If you have any concerns about our products, you can contact us on
ProductSafety@springernature.com

In case Publisher is established outside the EU, the EU authorized representative is:
Springer Nature Customer Service Center GmbH
Europaplatz 3, 69115 Heidelberg, Germany

Printed by Libri Plureos GmbH
in Hamburg, Germany